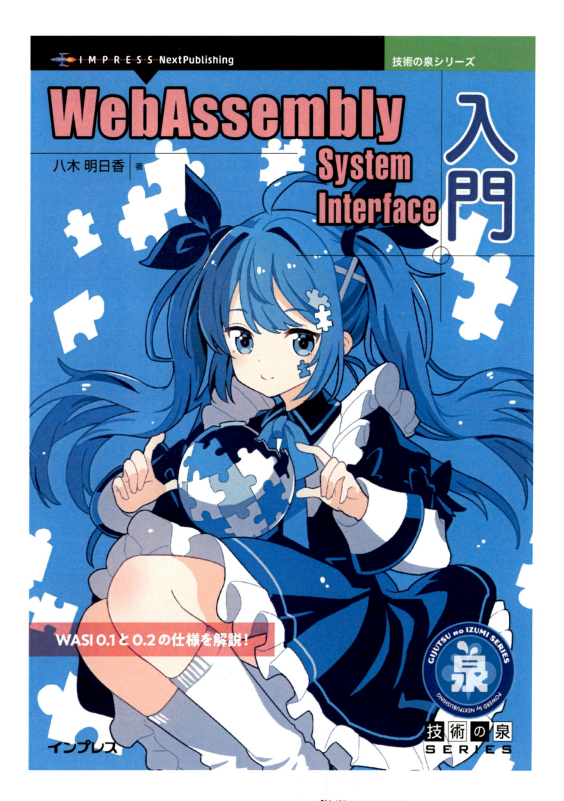

目次

はじめに ……………………………………………………………………………… 4

本書の構成 ………………………………………………………………………… 4

サンプルコードについて ………………………………………………………… 5

表記関係について ………………………………………………………………… 5

第1章　前提知識 …………………………………………………………………… 7

1.1　コマンドツール ……………………………………………………………… 7

1.2　WebAssembly …………………………………………………………………… 8

1.3　セキュリティー(安全性) ……………………………………………………… 10

1.4　ポータビリティー(移植性・柔軟性) ………………………………………… 14

1.5　WebAssembly が注目されている理由 ……………………………………… 16

1.6　ウェブ以外の環境で実行するために ……………………………………… 16

1.7　WASI(WebAssembly System Interface) …………………………………… 19

1.8　まとめ …………………………………………………………………………… 21

第2章　WASI 0.1 …………………………………………………………………… 23

2.1　WASI 0.1(プレビュー1) ……………………………………………………… 23

2.2　モジュールの種類 …………………………………………………………… 24

2.3　プロセスを終了する ………………………………………………………… 26

2.4　標準入力と標準出力 ………………………………………………………… 27

2.5　環境変数と引数 ……………………………………………………………… 34

2.6　現在時刻(UNIXTIME)を表示する …………………………………………… 40

2.7　乱数を表示する ……………………………………………………………… 43

2.8　ファイルの内容を表示する ………………………………………………… 46

2.9　ソケット通信 ………………………………………………………………… 53

2.10　WASI 0.1 の課題 …………………………………………………………… 56

2.11　まとめ ………………………………………………………………………… 57

2　目次

第3章　コンポーネントモデル ·· 59

3.1　コンポーネントモデル ··· 59

3.2　コンポーネントモデルの生まれた背景 ··· 59

3.3　WIT ·· 60

3.4　example:hello パッケージの実装 ··· 63

3.5　example:component パッケージの実装 ·· 65

3.6　まとめ ··· 68

第4章　WASI 0.2 ··· 69

4.1　WASI 0.2(プレビュー2) ··· 69

4.2　wasi:cli パッケージ ·· 70

4.3　wasi:io パッケージ ··· 90

4.4　wasi:clocks パッケージ ·· 92

4.5　wasi:random パッケージ ··· 96

4.6　wasi:filesystem パッケージ ·· 100

4.7　wasi:sockets パッケージ ··· 107

4.8　wasi:http パッケージ ··· 116

4.9　WASI 0.2 の課題 ·· 124

4.10　まとめ ··· 125

第5章　今後の展望 ··· 126

5.1　Warg ··· 126

5.2　WAC ··· 129

5.3　WIT 定義を公開する ··· 131

5.4　JCO ··· 132

5.5　まとめ ··· 133

付録A　APPENDIX ··· 134

A.1　参考文献 ··· 134

はじめに

　WASIはWasm(WebAssembly)のコミュニティーグループによって開発されている、Wasmのシステムインターフェースの仕様です。2023年8月にリリースされたGo 1.21からWASI 0.1へのビルドがサポート[1]されたり、翌年の2024年1月にWASI 0.2がリリース[2]されたりと、近年ますます注目されている技術です。

　WASIは、Wasmの標準的なシステムインターフェースを提供することを目的としています。本書がWASIの仕様を理解する手助けとなれば幸いです。

本書の構成

　本書は5つの章から構成されています。興味のある章から読み進めても問題ありません。以下、各章の章の概要を紹介します。

第1章 前提知識

WASIを解説するための前提知識を紹介します。Wasmについての簡単な解説と、WASIが生まれるに至った背景について解説します。また、本書で使用しているコマンドツールも第1章で紹介します。

第2章 WASI 0.1

WASI 0.1(プレビュー1)の基本的な仕様の解説と、WASI 0.1のインターフェースを用いた実装例を紹介します。2024年の1月にWASI 0.2がリリースされましたが、WASI 0.2が普及するにはもう少し時間がかかると思われます。実務でWASI 0.1を使用しているのであれば、本章が理解の助けになるでしょう。

第3章 コンポーネントモデル

WASI 0.2(プレビュー2)の仕様のベースとなるコンポーネントモデルについて解説します。コンポーネントモデルを採用したことによって、WASI 0.2はWASI 0.1とは全く異なるインターフェース仕様となりました。WASI 0.1はWasmの基本的な仕様の上に構築されていますが、WASI 0.2の仕様はコンポーネントモデルの仕様の上に構築されています。そのため、コンポーネントモデルのみを解説する章を設けています。第4章を読む前に、本章から読み進めることをお勧めします。

第4章 WASI 0.2

WASIの最新仕様であるWASI 0.2(プレビュー2)の基本的な仕様の解説と、WASI 0.2のインターフェースを用いた実装例を紹介します。WASI 0.1の実装例(第2章)と同様の実装例を紹介しているため、インターフェースがどのように変化したのか比較したい場合は第2章とあわせて読むことをお勧めします。

第5章 今後の展望

現在開発されているWasmのエコシステムについて紹介します。WasmのレジストリーであるWarg、Wasmコンポーネントを組み合わせるためのツールであるWACなどのエコシステム周りの開発が

1.Go 1.21 Release Notes:https://tip.golang.org/doc/go1.21

2.WASI 0.2 Launched:https://bytecodealliance.org/articles/WASI-0.2

進んでいます。Wargは現在パブリックベータ版が公開されているため、それらの使用方法についても紹介します。

サンプルコードについて

特に断りがない場合、本書で紹介しているコマンドおよびサンプルコードは全てLinuxもしくはUNIX(macOS)を前提に記述しています。また、サンプルコードはすべてGitHub上のリポジトリー[3]に公開しています。記載しているコードの断片から実装の全体像を読み取れない場合などは、ライセンスの範囲で自由にお使いください。

表記関係について

本書に記載されている会社名、製品名などは、一般に各社の登録商標または商標、商品名です。会社名、製品名については、本文中では©、®、™マークなどは表示していません。

---

3. サンプルコード:https://github.com/a-skua/wasi-book-example

第1章　前提知識

1.1　コマンドツール

本書で使用しているコマンドツールを簡単に紹介します。

Deno

Deno[1]は、TypeScriptとJavaScriptのランタイムです。公式サイトに記載されている下記のコマンドを実行することで、インストールすることができます。

```
$ curl -fsSL https://deno.land/install.sh | sh
...
```

Wasmtime

Wasmtime[2]は、WASI 0.1とWASI 0.2の両方をサポートしているWasmのランタイムです。GitHubのREADMEに記載されている下記のコマンドを実行することで、インストールすることができます。

```
$ curl https://wasmtime.dev/install.sh -sSf | bash
...
```

Rust

Rust[3]は、システムプログラミング言語です。公式サイトに記載されている下記のコマンドを実行することで、インストールすることができます。

```
$ curl --proto '=https' --tlsv1.2 -sSf https://sh.rustup.rs | sh
...
```

RustはWasmをビルドするためのエコシステムが整っているため、本書ではWasmの実装例にRustを使用しています。RustからWasmにビルドする場合、--targetオプションを用いてビルドターゲット(wasm32-unknown-unknownもしくはwasm32-wasip1)を指定します。これらのビルドター

1.https://deno.com/

2.https://github.com/bytecodealliance/wasmtime

3.https://www.rust-lang.org

ゲットがビルド環境のコンピューターに入っていない場合、次のコマンドでビルドターゲットを追加することができます。

```
$ rustup target add wasm32-unknown-unknown
...
```

```
$ rustup target add wasm32-wasip1
...
```

wasm-tools

wasm-tools[4]はWasmのバイナリーフォーマットとテキストフォーマットの相互変換や、Wasmの解析に用いるコマンドです。Rustのパッケージマネージャーであるcargoコマンドからインストールすることができます。

```
$ cargo install wasm-tools
...
```

cargo-component

cargo-component[5]は、WASI 0.2やWasmコンポーネントをビルドするのに使用するコマンドです。Rustのパッケージマネージャーであるcargoコマンドのサブコマンドとして提供されており、cargoコマンドからインストールすることができます。

```
$ cargo install cargo-component --locked
```

1.2　WebAssembly

Wasm (WebAssembly)は、ウェブブラウザー上で高速にプログラムを実行することを目的に開発された技術です。JavaScriptが人にとって理解しやすいテキストフォーマットであるのに対して、Wasmはコンピューターにとって理解しやすいバイナリーフォーマットとして設計されています。そのため、Wasmはコンピューター上で高速にプログラムを実行することができます。

4.https://github.com/bytecodealliance/wasm-tools

5.https://github.com/bytecodealliance/cargo-component

WAT

　バイナリーフォーマットがコンピューターにとって理解しやすい一方で、バイナリーフォーマットは人が読み書きするのに適していません。そのため、Wasmにはバイナリーフォーマットと相互変換可能なテキストフォーマットであるWAT(WebAssembly Text Format)[6]が用意されています。WATはS式と呼ばれる構文を採用しており、次のように書くことができます。

リスト1.1: WATのコード例(example.wat)

```
(module
  (func $add (param $a i32) (param $b i32) (result i32)
    local.get $a
    local.get $b
    i32.add
  )
  (export "add" (func $add))
)
```

WATを用いることでバイナリーフォーマットを扱いやすくはなりますが、それでもJavaScriptなどと比べると、まだまだ扱いにくい言語であることに変わりはありません。そのため、通常はRustなどの言語を用いてプログラムを書き、Wasmにコンパイルします。

リスト1.2: Rustのコード例(example/src/lib.rs)

```
#[no_mangle]
pub fn add(a: i32, b: i32) -> i32 {
    a + b
}
```

実行する

　Wasmは1ファイル1モジュールの構成になっており、モジュール単位でインスタンス化して実行します。Wasmモジュールを実行するには、モジュールファイルを読み込み、モジュールからインスタンスを作成する必要があります。ウェブ上でWasmを実行する場合、次のようなJavaScriptのコードを書く必要があります。

リスト1.3: JavaScriptのコード例(example.js)

```
// ファイル名
const filename = new URL("example.wasm", import.meta.url);

// バイナリーを取得
```

6.https://webassembly.github.io/spec/core/text/index.html

第1章　前提知識　9

```js
const binary = await fetch(filename).then((res) => res.arrayBuffer());

// モジュールを作成
const module = new WebAssembly.Module(binary);

// インスタンスを作成
const instance = new WebAssembly.Instance(module);

// 実行
console.log(instance.exports.add(1, 2)); // 3
```

このJavaScriptのコードをHTMLファイルに埋め込むか、Denoなどのランタイムを用いることで実行することができます。

```
$ deno run --allow-read example.js
3
```

1.3　セキュリティー(安全性)

Wasmにはウェブ上で高速にプログラムを実行できるという特徴以外にも、セキュリティー(安全性)とポータビリティー(移植性・柔軟性)にも注目すべき特徴があります。JavaScriptのコードを例に、Wasmのセキュリティーについて解説します。

JavaScriptから「Hello, world!」を出力する例

JavaScriptを用いて「Hello, world!」を標準出力(/dev/stdout)に出力するプログラムを実行する場合、次のようなコードを書きます。

リスト1.4: JavaScriptのコード例(hello.js)
```js
console.log("Hello, world!");
```

```
$ deno run hello.js
Hello, world!
```

このプログラムを実行すると、ターミナル上に「Hello, world!」と表示されます。この出力先である標準出力は、OSが管理しているリソースです。OSは管理しているリソースにアプリケーション(OS上で実行するプログラム)がアクセスできるように、インターフェース(API)を提供しています。このOSが提供しているインターフェースのことを、システムインターフェースと呼びます。

図 1.1: 標準出力「Hello, world!」を出力する例

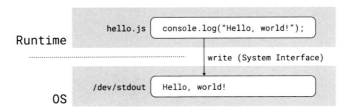

JavaScriptのランタイムは、`console.log`が実行されるとOSのシステムコール[7]を行い、標準出力に内容を出力します。

WebAssemblyから「Hello, world!」を出力する例

一方でWasmはサンドボックスとして設計されているため、OSのシステムインターフェースを直接実行することができません。そのため、Wasm単体では「Hello, world!」を標準出力に出力することができないようになっています。

Wasmから「Hello, world!」を出力するには、Wasmの外側の世界(JavaScriptのランタイム)に代わりに「Hello, world!」を出力してもらう必要があります。

Rustを用いて、「Hello, world!」を出力するWasmをビルドする場合、次のようなコードを書く[8]必要があります。

リスト 1.5: Rustのコード例(hello/src/lib.rs)

```rust
// 実行時にインポートするモジュールの定義
#[link(wasm_import_module = "env")]
extern "C" {
    fn write(ptr: *const u8, len: usize);
}

#[no_mangle]
pub fn _start() {
    let msg = b"Hello, world!";
    unsafe {
        write(msg.as_ptr(), msg.len());
    }
}
```

このコードからビルドしたWasmを実行するには、次のようなJavaScriptのコードを書く必要があります。

---

7. システムコール: システムインターフェースを実行すること
8. ここではアトミックなRustのコードを紹介していますが、Rustからより簡単にWasmを使えるようにするためのwasm-bindgenなどの便利なライブラリーが存在します。

リスト1.6: JavaScriptのコード例(hello/hello.js)

```
const filename = new URL(
  "target/wasm32-unknown-unknown/release/hello.wasm",
  import.meta.url,
);

// WebAssemblyの代わりに標準出力に出力する関数
function write(ptr, len) {
  console.log(new TextDecoder().decode(
    new Uint8Array(memory.buffer, ptr, len),
  ));
}

const { instance } = await WebAssembly.instantiateStreaming(
  fetch(filename),
  // モジュールのインポート
  { env: { write } },
);

const memory = instance.exports.memory;
instance.exports._start();
```

ビルドして実行すると、「Hello, world!」がターミナル上に表示されます。

```
$ cargo build --release --target wasm32-unknown-unknown
    Compiling hello v0.1.0 (/wasi-book-example/第1章/hello)
    Finished `release` profile [optimized] target(s) in 0.20s
```

```
$ deno run --allow-read hello.js
Hello, world!
```

　このようにWasmは直接OSのリソースにアクセスすることができないため、Rustのコード上で
Wasmのインスタンス作成時にインポートするenvモジュールを定義し、envモジュールのwrite関
数を呼び出しています。

　JavaScriptのコード上では、ポインターと文字列長を受け取り、Wasmのメモリー上から文字列
を読み込んでconsole.logを呼び出すwrite関数を実装しています。このwrite関数をインスタン
ス作成時にインポートするenvモジュールの関数として、Wasmに渡しています。

12 | 第1章 前提知識

図 1.2: WebAssembly から標準出力へ出力する例

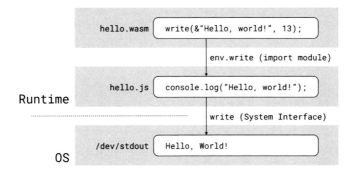

このようにすることで、Wasm は「Hello, world!」を標準出力に出力することができます。

WebAssembly の出力を制限する (1)

インスタンス作成時に write 関数をインポートしない場合はどうなるでしょうか?

リスト 1.7: write 関数をインポートしない例 (hello/hello.js)

```
const filename = new URL(
  "target/wasm32-unknown-unknown/release/hello.wasm",
  import.meta.url,
);

const { instance } = await WebAssembly.instantiateStreaming(
  fetch(filename),
  { env: {} },
);

instance.exports._start();
```

```
$ deno run --allow-read hello.js
error: Uncaught (in promise) LinkError: WebAssembly.instantiate(): Import #0
"env" "write": function import requires a callable
...
```

このように write 関数をインポートしないようコードを書き換えて実行すると、エラーとなり Wasm を実行することができません。

Wasm では、実行するランタイム側がアクセスさせたくない機能をインポートしないようにすることで、Wasm の実行を制限することができます。

WebAssemblyの出力を制限する(2)

　write関数をインポートしないようにするとWasm自体を実行できなくなるため、代わりに何も
しないwrite関数をインポートすることで機能を制限することもできます。

リスト1.8: write関数を空にする例(hello/hello.js)

```
const filename = new URL(
  "target/wasm32-unknown-unknown/release/hello.wasm",
  import.meta.url,
);

function write(_ptr, _len) {
  // DO NOTHING
}

const { instance } = await WebAssembly.instantiateStreaming(
  fetch(filename),
  { env: { write } },
);

instance.exports._start();
```

```
$ deno run --allow-read hello.js
```

　このようにWasmにはOSのリソースに直接アクセスする方法が存在しないため、Wasmを実行す
るランタイム側がWasmの挙動を簡単に制御することができます。

1.4　ポータビリティー(移植性・柔軟性)

　WasmのバイナリーフォーマットはIntelやARMなどのCPUアーキテクチャーや、Linuxや
WindowsなどのOSに依存していません。そのため、ランタイムさえあればCPUやOSを問わず、同
じバイナリーフォーマットのWasmを実行することができます。

14 　第1章　前提知識

図 1.3: CPU や OS に依存しないバイナリーフォーマット

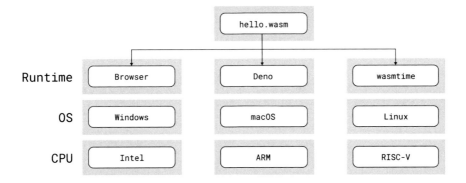

ウェブ上で実行する

ここまで Deno を用いて Wasm を実行する例を紹介しましたが、次のような HTML ファイルを用意することで、ウェブ上でも同じ Wasm を実行することができます。

リスト 1.9: HTML のコード例 (hello/hello.html)

```html
<!DOCTYPE html>
<html>
  <head>
    <meta charset="utf-8">
    <title>Example</title>
    <script type="module" src="hello.js"></script>
  </head>
  <body>
    <p>Open DevTools to see the console output.</p>
  </body>
</html>
```

```
$ python3 -m http.server
Serving HTTP on :: port 8000 (http://[::]:8000/) ...
...
```

図1.4: ウェブ上でWebAssemblyを実行する例

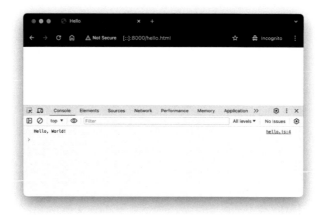

1.5　WebAssemblyが注目されている理由

ランタイムさえあればどこでも実行できるというのは、Wasmだけの特徴ではありません。Javaも同様に、ランタイムであるJVM(Java Virtual Machine)があれば、OSやCPUアーキテクチャーに依存することなくプログラムを実行することができます。

近年Wasmが注目されているのはOSやCPUに依存しないバイナリーフォーマットであることだけでなく、軽量で高速かつ安全に実行できるという特徴が揃っているからです。元々はウェブ上で高速にプログラムを実行することを目的に開発された技術でしたが、この特徴を活かすことでエッジコンピューティングのような限られたリソース上でプログラムを実行したり、Linuxコンテナの置き換えといった用途での活用が期待されている技術です。

1.6　ウェブ以外の環境で実行するために

Wasmから「Hello, world!」を出力するためには、Wasmの代わりに「Hello, world!」を出力するためのモジュールをインポートする必要がありました。

本書ではenvモジュールのwrite関数をインポートする実装例を紹介しましたが、別の人はconsoleモジュールのlog関数をインポートするWasmを実装するかもしれません。また、同じwrite関数でも、引数や戻り値が異なるかもしれません。

Wasmは、OSが管理しているリソースへのアクセス方法を定義していません。そのため、実装者によって必要なインポートモジュールやエクスポートする関数が異なる可能性があります。Wasmの実行者は、インポートしないといけないモジュールは何か、エクスポートされている関数は何かを知らないと、Wasmを実行することができません。

Rustの他にも、Wasmへのビルドをサポートしている言語にGoやDartなどがあります。これらの言語からWasmをビルドし、実行する方法を見てみましょう。

Goからビルドされた WebAssembly を実行する例

次のような Go のコードを用意します。

リスト 1.10: Go のコード例 (hello-go/hello.go)

```go
package main

import (
    "fmt"
)

func main() {
    fmt.Println("Hello, world!")
}
```

Go のコードを Wasm にビルドするには、次のコマンドを実行します。

```
$ env GOARCH=wasm GOOS=js go build -o hello.wasm hello.go
```

ビルドされた Wasm を実行するには、この Wasm のインターフェース(インポート/エクスポート)を知る必要があります。Go は Wasm を実行するための JavaScript のグルーコードを提供しています。次のコマンドを実行し、Go のグルーコードをコピーしましょう。

```
$ cp "$(go env GOROOT)/misc/wasm/wasm_exec.js" .
```

最後に、次のような JavaScript のコードを書くことで実行することができます。

リスト 1.11: JavaScript のコード例 (hello-go/hello.js)

```javascript
import "./wasm_exec.js";

const go = new Go();

const filename = new URL("./hello.wasm", import.meta.url);

const { instance } = await WebAssembly.instantiateStreaming(
  fetch(filename),
  go.importObject,
);

go.run(instance);
```

第1章 前提知識 | 17

```
$ deno run --allow-read hello.js
Hello, world!
```

Dartからビルドされた WebAssembly を実行する例

次のような Dart のコードを用意します。

リスト 1.12: Dart のコード例 (hello-dart/hello.dart)

```dart
void main() {
  print('Hello, world!');
}
```

Dart のコードを Wasm にビルドするには、次のコマンドを実行します。

```
$ dart compile wasm -o hello.wasm hello.dart
Generated wasm module 'hello.wasm', and JS init file 'hello.mjs'.
```

ビルドすると、hello.wasmと一緒にJavaScriptのグルーコード(hello.mjs)も出力されます。Dartの場合は次のような実行用のJavaScriptのコードを書くことで、Wasmを実行することができます。

リスト 1.13: JavaScript のコード例 (hello-dart/hello.js)

```javascript
import { instantiate, invoke } from "./hello.mjs";

const filename = new URL("./hello.wasm", import.meta.url);

const instance = await instantiate(
  WebAssembly.compileStreaming(fetch(filename)),
);

invoke(instance);
```

```
$ deno run --allow-read ./hello.js
Hello, world!
```

GoとDartのWebAssemblyの違い

GoとDartそれぞれからビルドしたWasmはどちらも同じ「Hello, world!」を出力するプログラムですが、提供されているインターフェースが異なるため、実行するのに必要なJavaScriptのコードも異なります。

18 | 第1章 前提知識

それぞれのWasmがどのようなモジュールをインポートしようとしているのかを確認すると、次のようになります。

```
$ wasm-tools print hello.wasm | grep '\(import '
  (import "dart2wasm" "_83" ...
  (import "dart2wasm" "_84" ...
  (import "dart2wasm" "_194" ...
...
```

```
$ wasm-tools print hello.wasm | grep '\(import '
  (import "gojs" "runtime.scheduleTimeoutEvent" ...
  (import "gojs" "runtime.clearTimeoutEvent" ...
  (import "gojs" "runtime.resetMemoryDataView" ...
...
```

DartからビルドされたWasmがdart2wasmモジュールをインポートする必要があるのに対して、GoからビルドされたWasmはgojsモジュールをインポートする必要あり、完全に異なるインターフェース仕様であることがわかります。

インターフェースの標準化

このようにビルドする環境によって必要なインターフェースが異なると、Wasmを実行するために必要なインターフェースを都度調べる必要があります。Goの例では、公式が用意しているグルーコードを使うことで実行できますが、Dartの場合はビルド時に一緒に生成されるグルーコードを使う必要があります。

Wasmは優れたポータビリティー性を持っていますが、標準化されたインターフェースが仕様に存在しないために、ビルド済みのWasmだけを他者に渡しても簡単に実行することができないという課題があります。この課題を解決するために、Wasmの標準化されたシステムインターフェースの仕様を策定しようという動きがあります。その標準化を目指しているWasmのシステムインターフェース仕様が、WASI(WebAssembly System Interface)です。

1.7 WASI(WebAssembly System Interface)

WASIを用いて、「Hello, world!」を出力するWasmを作成してみましょう。

GoからビルドされたWebAssemblyを実行する

先ほどのGoのコードをWASI 0.1にビルドして実行すると、次のようになります。

第1章　前提知識　19

リスト 1.14: Go のコード例 (hello-go/hello.go)

```go
package main

import (
    "fmt"
)

func main() {
    fmt.Println("Hello, world!")
}
```

```
$ env GOARCH=wasm GOOS=wasip1 go build -o hello.wasm hello.go
```

```
$ wasmtime run hello.wasm
Hello, world!
```

このように、Go からビルドされた Wasm を Wasmtime を使用して、簡単に実行することができます。

Rust からビルドされた WebAssembly を実行する

Dart は現状 WASI をサポートしていないため、代わりに WASI 0.1 をサポートしている Rust を使用して WASI 0.1 にビルドしてみましょう。

```
$ cargo new hello
...
```

リスト 1.15: Rust のコード例 (hello/src/main.rs)

```rust
fn main() {
    println!("Hello, world!");
}
```

Rust から WASI 0.1 に対応した Wasm をビルドするには、次のコマンドを実行します。

20 　第 1 章　前提知識

```
$ cargo build --release --target wasm32-wasip1
   Compiling hello v0.1.0 (/wasi-book-example/第1章/wasi/hello)
    Finished `release` profile [optimized] target(s) in 0.29s
```

Rustからビルドされた Wasm を実行するには、Go と同様に Wasmtime を使用します。

```
$ wasmtime run target/wasm32-wasip1/release/hello.wasm
Hello, World!
```

Go と Rust の WebAssembly の違い

　このように Go と Rust それぞれからビルドした Wasm に必要なインターフェースが何かを意識することなく、Wasmtime を使用するだけで簡単に実行することができました。これは Go と Rust が WASI 0.1 の仕様を満たした Wasm をビルドし、Wasmtime が WASI 0.1 をサポートしているためです。

　それぞれのインターフェースの違いは、次のようになります。

```
$ wasm-tools print  target/wasm32-wasip1/release/hello.wasm | grep '\(import '
  (import "wasi_snapshot_preview1" "fd_write" ...
  (import "wasi_snapshot_preview1" "environ_get" ...
  (import "wasi_snapshot_preview1" "environ_sizes_get" ...
  (import "wasi_snapshot_preview1" "proc_exit" ...
```

```
$ wasm-tools print hello.wasm | grep '\(import '
  (import "wasi_snapshot_preview1" "sched_yield" ...
  (import "wasi_snapshot_preview1" "proc_exit" ...
  (import "wasi_snapshot_preview1" "args_get" ...
...
```

　インポートしている関数の数が異なるものの、どちらも wasi_snapshot_preview1 モジュールをインポートしていることがわかります。この wasi_snapshot_preview1 モジュールが WASI 0.1 のモジュールです。

　このように共通のインターフェース仕様である WASI 0.1 を用いることで、どの言語からビルドされた Wasm かを意識することなく実行することができるようになります。

1.8　まとめ

　第1章では WASI の仕様を解説するための前提知識として、Wasm とはどういうものでどんな特

徴があるのかを紹介しました。また、WASIがなぜ必要とされているのか、WASIを使うとどういった メリットがあるのかを紹介しました。次章からは、WASIの仕様について解説していきます。

第2章 WASI 0.1

2.1 WASI 0.1(プレビュー1)

WASIの最初のバージョン[1]であるWASI 0.1(プレビュー1)は、POSIXとCloudABI[2]を参考にインターフェース仕様の策定が行われました。

WASI 0.1では、Wasmのコアスペック[3]をベースに、Wasmモジュールがエクスポートする内容とインポートするモジュールを定義しています。WASI 0.1のモジュールは`wasi_snapshot_preview1`モジュールをインポートし、エントリーポイントとなる`_start`関数をエクスポートするのが基本的な構成になります。

図2.1: WASI 0.1に準拠したWebAssemblyモジュールの基本構成

POSIXとCloudABI

POSIX (Portable Operating System Interface) は、OS (UNIX系) が提供するシステムインターフェースやライブラリー、コマンドラインなどのインターフェースを定めた標準仕様です。このPOSIXで定義されているコマンドだけを使用してシェルスクリプトを作成したり、定義されているAPIだけを使用してアプリケーションを作成することで、移植性の高いスクリプトやアプリケーションを作成することができます。

CloudABIは、Capability-based Securityのコンセプトに基づいて設計されたUNIX系OSのABIです。ケイパビリティ(Capability)は、OAuthのスコープと同様の概念です。どちらも最小権限の原則に基づいてユーザーに必要な権限を与え、権限に応じて操作を制御するという考え方を持ちます。残念ながらCloudABIは高い支持を得るに至らなかったため、現在プロジェクトはメンテナンスされていませんが、この考え方がWASIに取り入れられています。

WATで最小モジュールを用意する

WASI 0.1の仕様を満たす最小のWasmモジュールをWATで記述すると、次のようになります。

---

1. プレビュー0 (Unstable) と呼ばれる段階もありますが、ここでは無視しています
2. CloudABI:https://github.com/NuxiNL/cloudlibc
3. コアスペック:https://webassembly.github.io/spec/core/

リスト2.1: WATのコード例(command.wat)

```
(module
  (func $_start)
  (export "_start" (func $_start))
)
```

　このWasmモジュールには_start関数があり、その_start関数をエクスポートしています。Wasm
のランタイムはこの_start関数をエントリーポイントとして実行します。

　WATをWasmに変換するには、次のコマンドを実行します。

```
$ wasm-tools parse -o command.wasm command.wat
```

このWATの例では、_start関数に何も処理を書いていないため、何も実行せずにプログラムが終
了します。

```
$ wasmtime run command.wasm
```

2.2　モジュールの種類

　WASI 0.1には、コマンド(Command)モジュールとリアクター(Reactor)モジュールの2種類が用
意されています。先ほど紹介した、_start関数をエクスポートしているのが、実行用のコマンドモ
ジュールになります。

コマンドモジュール

　コマンドモジュールはプログラムの実行用モジュールです。デフォルトのエントリーポイントと
して、_start関数をエクスポートします。wasmtime runで実行されるのは、このコマンドモジュー
ルです。

リアクターモジュール

　コマンドモジュール以外のモジュール(実行時にインポートするモジュール)は、全てリアクター
モジュールです。リアクターモジュールはエントリーポイントとして、_initialize関数をエクス
ポートします。

　WATを用いて最小限のリアクターモジュールを書くと、次のようになります。

リスト 2.2: WAT のコード例 (reactor.wat)

```
(module
  (func $_init)
  (export "_initialize" (func $_init))
)
```

この_initialize関数はリアクターモジュールのエントリーポイントとして機能し、リアクター
モジュールをインスタンス化したときに、この_initialize関数が実行されます。

モジュールを組み合わせる

コマンドモジュールとリアクターモジュールは、組み合わせて実行することができます。次のよ
うなコマンドモジュールとリアクターモジュールを用意します。

リスト 2.3: コマンドモジュール (command.wat)

```
(module
  (import "reactor" "exitcode" (func $exitcode (result i32)))
  (memory $memory 1)
  (func $_start
    call $exitcode
    drop ;; スタックの内容を破棄
  )
  (export "_start" (func $_start))
  (export "memory" (memory $memory))
)
```

コマンドモジュールは、reactorモジュールのexitcode関数をインポートし、_start関数が実行
されると、exitcode関数を実行します。Wasmは最後に実行スタック内に残っている値の数と戻り
値の数が一致しないとエラーになるため、drop命令でスタックの内容を破棄しています。

リスト 2.4: リアクターモジュール (reactor.wat)

```
(module
  (global $EXITCODE (mut i32) (i32.const 0))
  (func $_init
    i32.const 1
    global.set $EXITCODE ;; グローバル変数に1をセット
  )
  (func $exitcode (result i32)
    global.get $EXITCODE
  )
  (export "_initialize" (func $_init))
  (export "exitcode" (func $exitcode))
```

第 2 章　WASI 0.1　25

リアクターモジュールは、_initialize関数が実行されると、グローバル変数$EXITCODEに1を
セットします。exitcode関数は、このグローバル変数の値を返します。

これらのWATをWasmに変換して実行してみましょう。Wasmtimeは--preloadオプションで
リアクターモジュールを指定することで、コマンドモジュールとリアクターモジュールを組み合わ
せて実行することができます。

```
$ wasm-tools parse -o command.wasm command.wat
```

```
$ wasm-tools parse -o reactor.wasm reactor.wat
```

```
$ wasmtime run --preload reactor=reactor.wasm command.wasm
```

何も表示されなければ、正常に実行されています。_start関数が実行される前に_initialize関
数が実行されていれば、exitcode関数の結果は1になっているはずです。

exitcode関数の結果を確認するために、exitcode関数の結果をWasmの終了コードとして、プ
ロセスを終了してみましょう。

2.3　プロセスを終了する

WASI 0.1には、プロセスを終了するためのproc_exit関数が用意されています。WASI 0.1の仕
様[4]には、次のように定義されています。

proc_exit(rval: exitcode)

Terminate the process normally.An exit code of 0 indicates successful termination of the
program.The meanings of other values is dependent on the environment.

また、exitcode型は次のように定義されています。

exitcode: u32

Exit code generated by a process when exiting.

このproc_exitの型定義をWATで表現すると、(func (param i32))となります。

先ほどのコマンドモジュールを、WASI 0.1のproc_exit関数を実行するように修正してみましょう。

4.WASI 0.1 の仕様:https://github.com/WebAssembly/WASI/blob/main/legacy/preview1/docs.md

リスト2.5: コマンドモジュール(command.wat)

```
(module
  (import "reactor" "exitcode" (func $exitcode (result i32)))
  (import "wasi_snapshot_preview1" "proc_exit" (func $proc_exit (param i32)))
  (memory $memory 1)
  (func $_start
    call $exitcode
    call $proc_exit
  )
  (export "_start" (func $_start))
  (export "memory" (memory $memory))
)
```

　新たにwasi_snapshot_preview1モジュールのproc_exit関数をインポートし、drop命令の代わりにproc_exit関数を実行しています。

　command.watを再度Wasmに変換し、実行してみましょう。

```
$ wasm-tools parse -o command.wasm command.wat
```

```
$ wasmtime run --preload reactor=reactor.wasm command.wasm
```

前回同様に何も表示されませんが、次のコマンドで終了コードが1になっていることを確認できます。

```
$ echo $?
1
```

2.4　標準入力と標準出力

　標準入力で名前を入力し、標準出力に「Hello, <YOUR NAME>!」と出力するプログラムをRustで書くと、次のようになります。

リスト2.6: Rustのコード例(stdio/src/main.rs)

```
use std::io::{stdin, stdout, Write};

fn main() {
    print!("Your name: ");
    stdout().flush().unwrap();
```

第2章　WASI 0.1　27

```rust
    // 標準入力
    let mut name = String::new();
    stdin().read_line(&mut name).unwrap();
    let name = name.trim();

    // 標準出力
    println!("Hello, {name}!");
}
```

このプログラムをWASI 0.1にビルドして実行すると、「Your name: 」と表示して入力待ち状態になり、「asuka」と入力すると、「Hello, asuka!」と出力されます。

```
$ cargo build --release --target wasm32-wasip1
    Compiling stdio v0.1.0 (/wasi-book-example/第2章/stdio)
     Finished `release` profile [optimized] target(s) in 1.11s
```

```
$ wasmtime run target/wasm32-wasip1/release/stdio.wasm
Your name: asuka
Hello, asuka!
```

Goでも、同様のコードをWASI 0.1にビルドして実行することができます。

リスト2.7: Goのコード例(stdio.go)

```go
package main

import (
    "fmt"
)

func main() {
    fmt.Print("Your name: ")

    // 標準入力
    var name string
    fmt.Scan(&name)

    // 標準出力
    fmt.Printf("Hello, %s!\n", name)
}
```

```
$ env GOOS=wasip1 GOARCH=wasm go build -o stdio.wasm stdio.go
```

```
$ wasmtime run stdio.wasm
Your name: asuka
Hello, asuka!
```

fd_read と fd_write

　先ほどの Rust と Go からビルドされた Wasm の内容を確認すると、fd_read 関数と fd_write 関数
がインポートされていることがわかります。

```
$ wasm-tools print target/wasm32-wasip1/release/stdio.wasm \
    | grep '\(import '
  (import "wasi_snapshot_preview1" "fd_read" ...
  (import "wasi_snapshot_preview1" "fd_write" ...
  (import "wasi_snapshot_preview1" "environ_get" ...
  (import "wasi_snapshot_preview1" "environ_sizes_get" ...
  (import "wasi_snapshot_preview1" "proc_exit" ...
```

```
$ wasm-tools print stdio.wasm | grep '\(import '
...
  (import "wasi_snapshot_preview1" "fd_read" ...
  (import "wasi_snapshot_preview1" "fd_write" ...
  (import "wasi_snapshot_preview1" "fd_fdstat_get" ...
  (import "wasi_snapshot_preview1" "fd_fdstat_set_flags" ...
  (import "wasi_snapshot_preview1" "fd_prestat_get" ...
  (import "wasi_snapshot_preview1" "fd_prestat_dir_name" ...
```

　この fd_read と fd_write が WASI 0.1 の入出力を扱う関数となります。仕様は次のように定義さ
れており、POSIX の readv 関数と writev 関数に似ていると記載されています。

fd_read(fd: fd, iovs: iovec_array) -> Result<size, errno>

Read from a file descriptor. Note: This is similar to readv in POSIX.

fd_write(fd: fd, iovs: ciovec_array) -> Result<size, errno>

Write to a file descriptor. Note: This is similar to writev in POSIX. Like POSIX, any calls of write (and other functions to read or write) for a regular file by other threads in the WASI process should not be interleaved while write is executed.

POSIXのreadvとwritev用いた実装例

　POSIXのreadv関数とwritev関数を使い、RustやGoで実装したコードと同様のコードを実装してみましょう。実装にはC言語を用います。

リスト2.8: C言語のコード例(stdio.c)

```c
#include <sys/uio.h>

int main() {
    char name[100];
    struct iovec iov[4];
    iov[0].iov_base = "Your name: ";
    iov[0].iov_len = 11;
    iov[1].iov_base = "Hello, ";
    iov[1].iov_len = 7;
    iov[2].iov_base = name;
    iov[2].iov_len = 100;
    iov[3].iov_base = "!\n";
    iov[3].iov_len = 2;

    writev(1, iov, 1);

    // 標準入力
    int len = readv(0, iov + 2, 1);
    iov[2].iov_len = len - 1; // 改行文字を含めない

    // 標準出力
    writev(1, iov + 1, 3);
    return 0;
}
```

readv関数とwritev関数は次のように定義されており、どちらも構造体iovecを用いてバッファーとバッファーサイズを指定して入出力を行います。

リスト2.9: readvとwritevの定義

```c
struct iovec {
    void *  iov_base;
    size_t  iov_len;
};

ssize_t readv(int d, const struct iovec *iov, int iovcnt);
```

30 | 第2章　WASI 0.1

```
ssize_t writev(int fildes, const struct iovec *iov, int iovcnt);
```

このコードをビルドして実行すると、RustやGoで実装したプログラムと同じ結果を得ることができます。

```
$ clang -o stdio.exe stdio.c
```

```
$ ./stdio.exe
Your name: asuka
Hello, asuka!
```

WASI 0.1のfd_readとfd_writeを用いた実装

C言語のコードを参考に、fd_read関数とfd_write関数を用いてWasmを作成してみましょう。Rustのプロジェクトを作成し、src/lib.rsの内容を変更していきます。

```
$ cargo new --lib fd_read_write
...
```

fd_read関数とfd_write関数を定義すると、次のようになります。

リスト2.10: fd_readとfd_writeの定義 (fd_read_write/src/lib.rs)

```
type Size = usize;
type Fd = usize;
type Errno = usize;
type Exitcode = usize;

#[repr(C)]
pub struct Iovec {
    pub buf: *mut u8,
    pub buf_len: Size,
}

#[repr(C)]
pub struct Ciovec {
    pub buf: *const u8,
    pub buf_len: Size,
}
```

```rust
#[link(wasm_import_module = "wasi_snapshot_preview1")]
extern "C" {
    pub fn fd_read(fd: Fd, iovs: *const Iovec, iovs_len: Size, read: *mut Size) -> Errno;
    pub fn fd_write(fd: Fd, ciovs: *const Ciovec, ciovs_len: Size, written: *mut Size) -> Errno;
    pub fn proc_exit(rval: Exitcode);
}
```

仕様書に定義されているfd_read(fd: fd, iovs: iovec_array) -> Result<size, errno>をWATで表現すると、(func (param i32 i32 i32 i32) (result i32))になります。このWATの表現をRustの型として定義すると、fn fd_read(fd: Fd, iovs: *const Iovec, iovs_len: Size, nread: *mut Size) -> Errnoとなります。

図2.2: fd_readの定義と実装

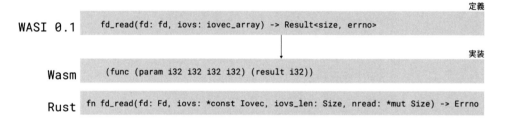

この定義を使用して、コマンドモジュールのエントリーポイントとなる_start関数を実装すると、次のようになります。

リスト2.11: fd_readとfd_writeを用いた実装例(fd_read_write/src/lib.rs)
```rust
#[no_mangle]
pub fn _start() {
    let mut nwritten: Size = 0;
    let mut nread: Size = 0;
    let mut buf = [0; 100];

    let ciovs = [Ciovec {
        buf: "Your name: ".as_ptr(),
        buf_len: 11,
    }];
    unsafe { fd_write(1, ciovs.as_ptr(), 1, &mut nwritten) };

    // 標準入力
    let iovs = [Iovec {
```

```rust
        buf: buf.as_mut_ptr(),
        buf_len: buf.len(),
    }];
    unsafe { fd_read(0, iovs.as_ptr(), 1, &mut nread) };

    // 標準出力
    let ciovs = [
        Ciovec {
            buf: "Hello, ".as_ptr(),
            buf_len: 7,
        },
        Ciovec {
            buf: buf.as_ptr(),
            buf_len: nread - 1, // 改行を含めない
        },
        Ciovec {
            buf: "!\n".as_ptr(),
            buf_len: 2,
        },
    ];
    // unsafe { fd_write(1, ciovs.as_ptr(), ciovs.len(), &mut nwritten) };
    for i in 0..ciovs.len() {
        unsafe { fd_write(1, ciovs[i..].as_ptr(), 1, &mut nwritten) };
    }

    unsafe { proc_exit(0) };
}
```

`for i in 0..ciovs.len() { ... }`と 書 く 代 わ り に 、`fd_write(1, ciovs.as_ptr(),` `ciovs.len(), &mut nwritten);`と書くこともできます。ここではWasmtimeが最初の要素しか出力してくれないため、for文を用いたループ処理を書いています[5]。

　Cargo.tomlに`lib.crate-type = "cdylib"`を追加してWasmにビルドし、実行してみましょう。また、`profile.release.lto = true`も追加すると、余分な情報を省いた軽量なWasmをビルドすることができます。

リスト2.12: Cargo.toml の設定例 (fd_read_write/Cargo.toml)

```toml
[package]
name = "fd_read_write"
version = "0.1.0"
```

---

5. これは Wasmtime のバグです。Wasmer では問題なく動作します。

```
edition = "2021"

[lib]
crate-type = ["cdylib"]

[profile.release]
lto = true

[dependencies]
```

```
$ cargo build --release --target wasm32-unknown-unknown
   Compiling fd_read_write v0.1.0 (/wasi-book-example/第2章/fd_read_write)
    Finished `release` profile [optimized] target(s) in 0.48s
```

```
$ wasmtime run target/wasm32-unknown-unknown/release/fd_read_write.wasm
Your name: asuka
Hello, asuka!
```

　このようにRustやGoを使用して、WASI 0.1をターゲットにビルドしたときと同じ結果を得ることができました。

2.5　環境変数と引数

　環境変数と引数を取得するプログラムをRustやGoで、次のように書くことができます。

リスト2.13: Rustのコード例(env/src/main.rs)

```
use std::env::{args, var};

fn main() {
    // 環境変数
    let example = var("EXAMPLE").unwrap_or("".to_string());
    println!("EXAMPLE = {example}");

    // 引数
    let args: Vec<String> = args().collect();
    println!("ARGS = {args:?}");
}
```

```
$ cargo build --release --target wasm32-wasip1
   Compiling env v0.1.0 (/wasi-book-example/第2章/env)
    Finished `release` profile [optimized] target(s) in 0.24s
```

```
$ wasmtime run --env EXAMPLE=foo \
    ./target/wasm32-wasip1/release/env.wasm bar
EXAMPLE = foo
ARGS = ["env.wasm", "bar"]
```

リスト2.14: Goのコード例(env.go)

```go
package main

import (
    "fmt"
    "os"
)

func main() {
    // 環境変数
    env := os.Getenv("EXAMPLE")
    fmt.Println("EXAMPLE =", env)

    // 引数
    args := os.Args
    fmt.Println("ARGS =", args)
}
```

```
$ env GOOS=wasip1 GOARCH=wasm go build -o env.wasm env.go
```

```
$ wasmtime run --env EXAMPLE=foo ./env.wasm bar
EXAMPLE = foo
ARGS = [env.wasm bar]
```

第2章 WASI 0.1 | 35

RustとGoそれぞれからビルドしたWasmの内容を確認すると、environ_get関数とenviron_sizes_get関数、args_get関数とargs_sizes_get関数がインポートされていることがわかります。

```
$ wasm-tools print target/wasm32-wasip1/release/env.wasm | grep '\(import '
  (import "wasi_snapshot_preview1" "args_sizes_get" ...
  (import "wasi_snapshot_preview1" "args_get" ...
  (import "wasi_snapshot_preview1" "fd_write" ...
  (import "wasi_snapshot_preview1" "environ_get" ...
  (import "wasi_snapshot_preview1" "environ_sizes_get" ...
  (import "wasi_snapshot_preview1" "proc_exit" ...
```

```
$ wasm-tools print ./env.wasm | grep '\(import '
...
  (import "wasi_snapshot_preview1" "args_get" ...
  (import "wasi_snapshot_preview1" "args_sizes_get" ...
...
  (import "wasi_snapshot_preview1" "environ_get" ...
  (import "wasi_snapshot_preview1" "environ_sizes_get" ...
...
```

environ_getとenviron_sizes_get

environ_getとenviron_sizes_getは環境変数を取得するための関数で、仕様は次のとおりです。

environ_get(environ: Pointer\<Pointer\<u8\>\>, environ_buf: Pointer\<u8\>) -> Result\<(), errno\>

Read environment variable data. The sizes of the buffers should match that returned byenviron_sizes_get. Key/value pairs are expected to be joined with=s, and terminated with\0s.

environ_sizes_get() -> Result\<(size, size), errno\>

Return environment variable data sizes.

environ_sizes_getの戻り値は(size, size)となっており、それぞれ環境変数の数と環境変数のデータサイズを表します。

environ_getはenviron_sizes_getで取得した環境変数分のポインターと、サイズ分のバッファーを用意して実行します。ポインター列には各環境変数の先頭ポインターが格納され、バッファーにはキーと値が=で結合された状態で格納されます。環境変数の終了を表すために、\0(NULL文字)が使用されます。

たとえばFOO=1、BAR=2、BAZ=3の環境変数がある場合、図のようなデータ構造になります。

図2.3: 環境変数を取得するときのデータ構造

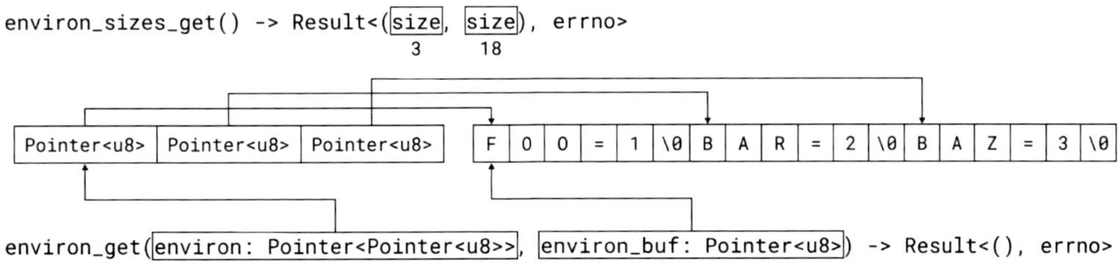

```
environ_sizes_get() -> Result<(size, size), errno>
                              3     18
```

```
| Pointer<u8> | Pointer<u8> | Pointer<u8> |   | F | O | O | = | 1 | \0 | B | A | R | = | 2 | \0 | B | A | Z | = | 3 | \0 |
```

```
environ_get(environ: Pointer<Pointer<u8>>, environ_buf: Pointer<u8>) -> Result<(), errno>
```

　Rustのプロジェクトを作成し、src/lib.rsの内容を変更していきましょう。environ_getと environ_sizes_get関数を用いて環境変数を取得するプログラムをRustで書くと、次のように なります。

```
$ cargo new --lib environ_get
...
```

リスト2.15: 環境変数を取得する実装例(environ_get/src/lib.rs)
```
...
#[link(wasm_import_module = "wasi_snapshot_preview1")]
extern "C" {
    pub fn environ_get(environ: *mut usize, environ_buf: *mut u8) -> Errno;
    pub fn environ_sizes_get(environ_count: *mut Size, environ_buf_size: *mut
Size) -> Errno;
}

#[no_mangle]
pub fn _start() {
    let mut environ_count: Size = 0;
    let mut environ_buf_size: Size = 0;
    unsafe { environ_sizes_get(&mut environ_count, &mut environ_buf_size) };

    let mut environ = vec![0; environ_count];
    let mut environ_buf = vec![0; environ_buf_size];
    unsafe { environ_get(environ.as_mut_ptr(), environ_buf.as_mut_ptr()) };
...
}
```

環境変数を表示する部分はfd_writeを用いて、次のように実装します。

第2章　WASI 0.1　37

リスト2.16: 環境変数を表示する実装例(environ_get/src/lib.rs)

```rust
...
#[no_mangle]
pub fn _start() {
...
    for environ in environ {
        let environ = environ - environ_buf.as_ptr() as usize;
        let environ_buf = &environ_buf[environ..];
        let environ_buf = &environ_buf[..environ_buf.iter().position(|&byte| byte
== 0).unwrap()];
        let environ = std::str::from_utf8(environ_buf).unwrap();
        let environ: Vec<&str> = environ.split('=').collect();
        let ciovs = [
            Ciovec {
                buf: environ[0].as_ptr(),
                buf_len: environ[0].len(),
            },
            Ciovec {
                buf: " = ".as_ptr(),
                buf_len: 3,
            },
            Ciovec {
                buf: environ[1].as_ptr(),
                buf_len: environ[1].len(),
            },
            Ciovec {
                buf: "\n".as_ptr(),
                buf_len: 1,
            },
        ];
        // unsafe { fd_write(1, ciovs.as_ptr(), ciovs.len(), &mut 0) };
        for i in 0..ciovs.len() {
            unsafe { fd_write(1, ciovs[i..].as_ptr(), 1, &mut 0) };
        }
    }

    unsafe { proc_exit(0) };
}
```

Cargo.tomlに`lib.crate-type = "cdylib"`を追加してWasmにビルドし、実行してみましょう。

```
$ cargo build --release --target wasm32-unknown-unknown
    Compiling environ_get v0.1.0 (/wasi-book-example/第2章/environ_get)
    Finished `release` profile [optimized] target(s) in 0.15s
```

```
$ wasmtime run --env FOO=1 --env BAR=2 --env BAZ=3 \
    target/wasm32-unknown-unknown/release/environ_get.wasm
FOO = 1
BAR = 2
BAZ = 3
```

args_getとargs_sizes_get

args_getとargs_sizes_getは、引数を取得するための関数です。

args_get(argv: Pointer<Pointer<u8>>, argv_buf: Pointer<u8>) -> Result<(), errno>

Read command-line argument data. The size of the array should match that returned by args_sizes_get. Each argument is expected to be \0 terminated.

args_sizes_get() -> Result<(size, size), errno>

Return command-line argument data sizes.

args_getとargs_sizes_getのデータ構造はenviron_getとenviron_sizes_getと同じです。引数foo、bar、bazを渡したときのデータ構造は、図のようになります。

図2.4: 引数を取得するときのデータ構造

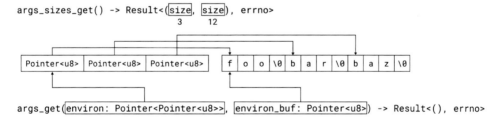

Rustのプロジェクトを作成し、src/lib.rsの内容を変更していきましょう。args_getとargs_sizes_get関数を用いて引数を取得するプログラムをRustで書くと、次のようになります。

```
$ cargo new --lib args_get
...
```

リスト2.17: 引数を取得する実装例(args_get/src/lib.rs)

```
...
#[link(wasm_import_module = "wasi_snapshot_preview1")]
extern "C" {
    pub fn args_get(argv: *mut usize, argv_buf: *mut u8) -> Errno;
    pub fn args_sizes_get(argv_count: *mut Size, argv_buf_size: *mut Size) ->
Errno;
}


#[no_mangle]
pub fn _start() {
    let mut argv_count: Size = 0;
    let mut argv_buf_size: Size = 0;
    unsafe { args_sizes_get(&mut argv_count, &mut argv_buf_size) };

    let mut argv = vec![0; argv_count];
    let mut argv_buf = vec![0; argv_buf_size];
    unsafe { args_get(argv.as_mut_ptr(), argv_buf.as_mut_ptr()) };
...
}
```

Cargo.tomlにlib.crate-type = "cdylib"を追加してWasmにビルドし、実行してみましょう。

```
$ cargo build --release --target wasm32-unknown-unknown
    Compiling args_get v0.1.0 (/wasi-book-example/第2章/args_get)
     Finished `release` profile [optimized] target(s) in 0.15s
```

```
$ wasmtime run target/wasm32-unknown-unknown/release/args_get.wasm foo bar baz
ARGS[0] = args_get.wasm
ARGS[1] = foo
ARGS[2] = bar
ARGS[3] = baz
```

2.6　現在時刻(UNIXTIME)を表示する

現在時刻(UNIXTIME)を表示するプログラムは、RustやGoで次のように書くことができます。

リスト2.18: Rustのコード例(time/src/main.rs)

```rust
use std::time::SystemTime;

fn main() {
    let now = SystemTime::now()
        .duration_since(SystemTime::UNIX_EPOCH)
        .unwrap();
    let now = now.as_secs();
    println!("{now}");
}
```

```
$ cargo build --release --target wasm32-wasip1
   Compiling time v0.1.0 (/wasi-book-example/第2章/time)
    Finished `release` profile [optimized] target(s) in 0.09s
```

```
$ wasmtime run target/wasm32-wasip1/release/time.wasm
1735052400
```

リスト2.19: Goのコード例(time.go)

```go
package main

import (
    "time"
    "fmt"
)

func main() {
    t := time.Now()
    fmt.Println(t.Unix())
}
```

```
$ env GOOS=wasip1 GOARCH=wasm go build -o time.wasm time.go
```

第2章 WASI 0.1 | 41

```
$ wasmtime run time.wasm
1735052400
```

この時刻を取得するのに使われているのが、WASI 0.1のclock_time_get関数です。

```
$ wasm-tools print target/wasm32-wasip1/release/time.wasm | grep '\(import '
  (import "wasi_snapshot_preview1" "clock_time_get" ...
  (import "wasi_snapshot_preview1" "fd_write" ...
  (import "wasi_snapshot_preview1" "environ_get" ...
  (import "wasi_snapshot_preview1" "environ_sizes_get" ...
  (import "wasi_snapshot_preview1" "proc_exit" ...
```

```
$ wasm-tools print time.wasm | grep '\(import '
...
  (import "wasi_snapshot_preview1" "clock_time_get" ...
...
```

clock_time_get関数は、次のように定義されています。

clock_time_get(id: clockid, precision: timestamp) -> Result<timestamp, errno>

Return the time value of a clock. Note: This is similar toclock_gettimein POSIX.

WASI 0.1では、timestampをns(ナノ秒)単位で表現します。

Rustのプロジェクトを作成し、src/lib.rsの内容を変更していきましょう。clock_time_get関数を用いて現在時刻を表示するプログラムをRustで書くと、次のようになります。

```
$ cargo new --lib clock_time_get
...
```

リスト2.20: 時刻を表示する実装例(clock_time_get/src/lib.rs)

```
...
#[link(wasm_import_module = "wasi_snapshot_preview1")]
extern "C" {
    pub fn clock_time_get(id: Clockid, precision: Timestamp, time: *mut
Timestamp) -> Errno;
    pub fn fd_write(fd: Fd, ciovs: *const Ciovec, ciovs_len: Size, written: *mut
Size) -> Errno;
    pub fn proc_exit(rval: Exitcode);
}
```

42 | 第2章 WASI 0.1

```rust
#[no_mangle]
pub fn _start() {
    let mut time: Timestamp = 0;
    unsafe { clock_time_get(0, 0, &mut time) };

    // ナノ秒 --> 秒
    let time = (time / 1_000_000_000).to_string();

    let ciovs = [
        Ciovec {
            buf: time.as_ptr(),
            buf_len: time.len(),
        },
        Ciovec {
            buf: "\n".as_ptr(),
            buf_len: 1,
        },
    ];
    // unsafe { fd_write(1, ciovs.as_ptr(), ciovs.len(), &mut 0) };
    for i in 0..ciovs.len() {
        unsafe { fd_write(1, ciovs.as_ptr().add(i), 1, &mut 0) };
    }
    unsafe { proc_exit(0) };
}
```

Cargo.tomlに`lib.crate-type = "cdylib"`を追加してWasmにビルドし、実行してみましょう。

```
$ cargo build --release --target wasm32-unknown-unknown
   Compiling clock_time_get v0.1.0 (/wasi-book-example/第2章/clock_time_get)
    Finished `release` profile [optimized] target(s) in 0.27s
```

```
$ wasmtime run target/wasm32-unknown-unknown/release/clock_time_get.wasm
1735052400
```

2.7 乱数を表示する

乱数を表示するプログラムをRustやGoで、次のように書くことができます。

リスト 2.21: Rust のコード例 (random/src/main.rs)

```rust
use rand::prelude::*;

fn main() {
    let num: u64 = random();
    println!("{num}");
}
```

```
$ cargo build --release --target wasm32-wasip1
...
    Compiling random v0.1.0 (/wasi-book-example/第2章/random)
     Finished `release` profile [optimized] target(s) in 1.42s
```

```
$ wasmtime run target/wasm32-wasip1/release/random.wasm
16160081518925629637
```

リスト 2.22: Go のコード例 (random.go)

```go
package main

import (
    "fmt"
    "math/rand"
)

func main() {
    num := rand.Uint64()
    fmt.Println(num)
}
```

```
$ env GOOS=wasip1 GOARCH=wasm go build -o random.wasm random.go
```

```
$ wasmtime run random.wasm
794042580185812991
```

WASI 0.1で乱数を生成するのに使われているのが、random_get関数です。

```
$ wasm-tools print target/wasm32-wasip1/release/random.wasm | grep '\(import '
  (import "wasi_snapshot_preview1" "random_get" ...
  (import "wasi_snapshot_preview1" "fd_write" ...
  (import "wasi_snapshot_preview1" "environ_get" ...
  (import "wasi_snapshot_preview1" "environ_sizes_get" ...
  (import "wasi_snapshot_preview1" "proc_exit" ...
```

```
$ wasm-tools print random.wasm | grep '\(import '
...
  (import "wasi_snapshot_preview1" "random_get" (func (;8;) (type 3)))
...
```

random_get関数は、次のように定義されています。

random_get(buf: Pointer<u8>, buf_len: size) -> Result<(), errno>

Write high-quality random data into a buffer. This function blocks when the implementation is unable to immediately provide sufficient high-quality random data. This function may execute slowly, so when large mounts of random data are required, it's advisable to use this function to seed a pseudo-random number generator, rather than to provide the random data directly.

Rustのプロジェクトを作成し、src/lib.rsの内容を変更していきましょう。random_get関数を用いて乱数を表示するプログラムをRustで書くと、次のようになります。

```
$ cargo new --lib random_get
...
```

リスト2.23: 乱数を表示する実装例(random_get/src/lib.rs)

```
...
#[link(wasm_import_module = "wasi_snapshot_preview1")]
extern "C" {
    pub fn random_get(buf: *mut u8, buf_len: Size) -> Errno;
    pub fn fd_write(fd: Fd, ciovs: *const Ciovec, ciovs_len: Size, written: *mut
Size) -> Errno;
    pub fn proc_exit(rval: Exitcode);
}

#[no_mangle]
pub fn _start() {
```

第2章　WASI 0.1 | 45

```rust
    let mut buf = [0; 8];
    unsafe { random_get(buf.as_mut_ptr(), buf.len()) };


    let num = u64::from_ne_bytes(buf).to_string();
    let ciovs = [
        Ciovec {
            buf: num.as_ptr(),
            buf_len: num.len(),
        },
        Ciovec {
            buf: "\n".as_ptr(),
            buf_len: 1,
        },
    ];
    // unsafe { fd_write(1, ciovs.as_ptr(), ciovs.len(), &mut 0) };
    for i in 0..ciovs.len() {
        unsafe { fd_write(1, ciovs.as_ptr().add(i), 1, &mut 0) };
    }


    unsafe { proc_exit(0) };
}
```

Cargo.tomlに`lib.crate-type = "cdylib"`を追加してWasmにビルドし、実行してみましょう。

```
$ cargo build --release --target wasm32-unknown-unknown
   Compiling random_get v0.1.0 (/wasi-book-example/第2章/random_get)
    Finished `release` profile [optimized] target(s) in 0.28s
```

```
$ wasmtime run target/wasm32-unknown-unknown/release/random_get.wasm
12524696322301836042
```

2.8　ファイルの内容を表示する

ファイルの内容を出力するプログラムをRustやGoで、次のように書くことができます。

リスト2.24: Rustのコード例(file/src/main.rs)

```rust
use std::env::args;
use std::fs::File;
use std::io::Read;

fn main() {
    let path = args().nth(1).unwrap();
    let mut file = File::open(path).unwrap();
    let mut contents = String::new();
    file.read_to_string(&mut contents).unwrap();

    println!("{contents}");
}
```

```
$ cargo build --release --target wasm32-wasip1
    Compiling file v0.1.0 (/wasi-book-example/第2章/file)
     Finished `release` profile [optimized] target(s) in 0.14s
```

```
$ wasmtime run --dir .::/ target/wasm32-wasip1/release/file.wasm /Cargo.toml
[package]
name = "file"
version = "0.1.0"
edition = "2021"
...
```

リスト2.25: Goのコード例(file.go)

```go
package main

import (
    "fmt"
    "io"
    "os"
)

func main() {
    path := os.Args[1]
    file, err := os.Open(path)
    if err != nil {
```

第2章 WASI 0.1 47

```
        panic(err)
    }
    defer file.Close()

    contents, err := io.ReadAll(file)
    if err != nil {
        panic(err)
    }

    fmt.Println(string(contents))
}
```

```
$ env GOOS=wasip1 GOARCH=wasm go build -o file.wasm file.go
```

```
$ wasmtime run --dir .::/ file.wasm /file.go
package main

import (
    "fmt"
    "io"
    "os"
)
...
```

WASIのファイルアクセス

WASIは、事前にアクセスを許可していないファイルにアクセスできないよう設計されています。

Wasmtimeの--dirオプションは、Wasmがアクセス可能なディレクトリーをマッピングするために使用します。--dirオプションの書式は--dir <HOST_DIR[::GUEST_DIR]>で、::区切りでホストディレクトリー(実際のOS上のディレクトリー)とゲストディレクトリー(Wawmから見たディレクトリー)を指定することができます。

ファイルの読み出しは標準入力でも使用したfd_read関数を使いますが、この関数を使うためにはファイルディスクリプター(FD)を知る必要があります。

POSIXの世界でファイルディスクリプターは「0」が標準入力、「1」が標準出力、「2」が標準エラー出力としてあらかじめ割り当てられています。WASI 0.1はPOSIXを参考に策定されているため、POSIXと同様の設計になっています。

POSIXではファイルを開くことで新たなファイルディスクリプターを得ることができますが、

WASIは事前にアクセスが許可されているファイルディスクリプター[6]にしかアクセスすることができません。

「0」、「1」、「2」以外のファイルディスクリプターに割り当てがあるかを知るために用いるのがfd_prestat_get関数です。

fd_prestat_get

fd_prestat_get関数はファイルディスクリプターに割り当てられているディレクトリーの情報を取得するための関数で、仕様は次のとおりです。

fd_prestat_get(fd: fd) -> Result<prestat, errno>

Return a description of the given preopened file descriptor.

WASI 0.1では事前にアクセスが許可されているファイルディスクリプターは、「3」から順に割り当てられます。RustやGoからビルドされたWasmは、fd_prestat_get関数を使い「3」以降のファイルディスクリプターに割り当てがあるかを順に探索する処理が実行時に走るようになっています。

Rustのプロジェクトを作成し、src/lib.rsの内容を変更していきましょう。Rustで事前にアクセスが許可されているファイルディスクリプターのディレクトリー名を表示するプログラムを書くと、次のようになります。

```
$ cargo new --lib fd_prestat_get
...
```

リスト 2.26: ファイルディスクリプターのディレクトリー名を表示する実装例 (fd_prestat_get/src/lib.rs)

```
...
#[link(wasm_import_module = "wasi_snapshot_preview1")]
extern "C" {
    pub fn fd_prestat_get(fd: Fd, prestat: *mut Prestat) -> Errno;
    pub fn fd_prestat_dir_name(fd: Fd, path: *mut u8, path_len: Size) -> Errno;
    pub fn fd_write(fd: Fd, ciovs: *const Ciovec, ciovs_len: Size, written: *mut
Size) -> Errno;
    pub fn proc_exit(rval: Exitcode);
}


#[no_mangle]
pub fn _start() {
    let mut fd = 3;
    let mut prestat = Prestat {
        tag_size: 0,
```

---

6. 事前にアクセスが許可されているというのは、すでに開かれているファイルディスクリプターであるため、仕様書では preopend というワードで紹介されています。

```rust
        _pad: [0; 3],
        prestat_dir: PrestatDir { pr_name_len: 0 },
    };

    while let 0 = unsafe { fd_prestat_get(fd, &mut prestat) } {
        let mut path = vec![0; prestat.prestat_dir.pr_name_len];
        unsafe { fd_prestat_dir_name(fd, path.as_mut_ptr(), path.len()) };

        let ciovs = [
            Ciovec {
                buf: "preopened: ".as_ptr(),
                buf_len: 11,
            },
            Ciovec {
                buf: path.as_ptr(),
                buf_len: path.len(),
            },
            Ciovec {
                buf: "\n".as_ptr(),
                buf_len: 1,
            },
        ];
        unsafe { fd_write(1, ciovs.as_ptr(), ciovs.len(), &mut 0) };
        for i in 0..ciovs.len() {
            unsafe { fd_write(1, ciovs.as_ptr().add(i), 1, &mut 0) };
        }
        fd += 1;
    }

    unsafe { proc_exit(0) };
}
```

　ファイルディスクリプターはそれがディレクトリーなのか、ファイルなのかを区別しませんが、事前にアクセスが許可されているファイルディスクリプターは、ディレクトリーであることを前提としています。

　fd_prestat_get関数から取得できる情報には、ファイルディスクリプターのディレクトリー名の文字列長しか含まれていません。そのため、fd_prestat_dir_name関数を用いてファイルディスクリプターのディレクトリー名を取得しています。

　Cargo.tomlにlib.crate-type = "cdylib"を追加してWasmにビルドし、実行してみましょう。--dir ./srcと--dir ./targetを指定して実行すると、それぞれのディレクトリー名が表示され

50　　第2章　WASI 0.1

ます。

```
$ cargo build --release --target wasm32-unknown-unknown
    Compiling fd_prestat_get v0.1.0 (/wasi-book-example/第2章/fd_prestat_get)
    Finished `release` profile [optimized] target(s) in 0.12s
```

```
$ wasmtime run --dir ./src --dir ./target \
    target/wasm32-unknown-unknown/release/fd_prestat_get.wasm
preopened: ./src
preopened: ./target
```

ファイルの内容を表示する

　ファイルの内容表示するには、ファイルを開く必要があります。ディレクトリーの中に含まれているファイルのファイルディスクリプターを取得するために用いる関数が、path_openです。

path_open

　path_open関数はファイルやディレクトリーを開くための関数で、仕様は次のとおりです。

path_open(fd: fd, dirflags: lookupflags, path: string, oflags: oflags, fs_rights_base: rights, fs_rights_inheriting: rights, fdflags: fdflags) -> Result<fd, errno>

Open a file or directory. The returned file descriptor is not guaranteed to be the lowest-numbered file descriptor not currently open; it is randomized to prevent applications from depending on making assumptions about indexes, since this is error-prone in multi-threaded contexts. The returned file descriptor is guaranteed to be less than $2^{**}31$. Note: This is similar toopenatin POSIX.

　path_open関数から得られたファイルディスクリプターをfd_read関数に渡すことで、ファイルの内容を読み出すことができます。

　Rustのプロジェクトを作成し、src/lib.rsの内容を変更していきましょう。path_open関数を用いてファイルを読み出すプログラムをRustで書くと、次のようになります。

```
$ cargo new --lib path_open
...
```

第2章　WASI 0.1　　51

リスト2.27: ファイルを読み出す実装例(path_open/src/lib.rs)

```rust
...
#[no_mangle]
pub fn _start() {
    let fd = 3;
    let mut prestat = Prestat {
        tag_size: 0,
        _pad: [0; 3],
        prestat_dir: PrestatDir { pr_name_len: 0 },
    };
    let errno = unsafe { fd_prestat_get(fd, &mut prestat) };
    if errno > 0 {
        panic!("fd_prestat_get failed: {errno}");
    }

    // 引数からファイル名を取得
...

    // ファイルを開く
    let mut new_fd = 0;
    let errno = unsafe { path_open(fd, 0, path.as_ptr(), path.len(), 0, 0, 0, 0,
&mut new_fd) };
    if errno > 0 {
        panic!("path_open failed: {errno}");
    }

    // ファイルを読み込む
    let mut nread: Size = 0;
    let mut buf = [0; 1024];
    let iovs = [Iovec {
        buf: buf.as_mut_ptr(),
        buf_len: buf.len(),
    }];
    unsafe { fd_read(new_fd, iovs.as_ptr(), iovs.len(), &mut nread) };

    // 標準出力
    let ciovs = [Ciovec {
        buf: buf.as_ptr(),
        buf_len: nread,
    }];
    unsafe { fd_write(1, ciovs.as_ptr(), ciovs.len(), &mut 0) };
```

52 | 第2章 WASI 0.1

```
    unsafe { proc_exit(0) };
}
```

Cargo.tomlに`lib.crate-type = "cdylib"`を追加してWasmにビルドし、実行してみましょう。

```
$ cargo build --release --target wasm32-unknown-unknown
   Compiling path_open v0.1.0 (/wasi-book-example/第2章/path_open)
    Finished `release` profile [optimized] target(s) in 0.26s
```

```
$ wasmtime run --dir . \
    target/wasm32-unknown-unknown/release/path_open.wasm Cargo.toml
[package]
name = "path_open"
version = "0.1.0"
edition = "2021"
...
```

2.9　ソケット通信

　WASI 0.1では新たにソケットを作成することはできませんが、事前に作成されたソケットにアクセスすることはできます。ソケットを扱う関数として`sock_accept`、`sock_recv`、`sock_send`、`sock_shutdown`の4つが用意されています。仕様は次のとおりです。

sock_accept(fd: fd, flags: fdflags) -> Result<fd, errno>

Accept a new incoming connection. Note: This is similar to accept in POSIX.

sock_recv(fd: fd, ri_data: iovec_array, ri_flags: riflags) -> Result<(size, roflags), errno>

Receive a message from a socket. Note: This is similar to recv in POSIX, though it also supports reading the data into multiple buffers in the manner of readv.

sock_send(fd: fd, si_data: ciovec_array, si_flags: siflags) -> Result<size, errno>

Send a message on a socket. Note: This is similar to send in POSIX, though it also supports writing the data from multiple buffers in the manner of writev.

sock_shutdown(fd: fd, how: sdflags) -> Result<(), errno>

Shut down socket send and receive channels. Note: This is similar to shutdown in POSIX.

HTTPサーバーを作成する

　これらの関数を使用して、HTTPリクエストを受け取り、「Hello, WASI 0.1!」というレスポンス

を返すプログラムを書く場合、次のような手順になります。

1. sock_accept: HTTPリクエストを待つ。
2. sock_send: HTTPレスポンスを返す。
3. sock_shutdown: ソケットを閉じる。

Rustのプロジェクトを作成し、src/lib.rsの内容を変更していきましょう。

```
$ cargo new --lib socket
...
```

リスト2.28: HTTPサーバーの実装例(socket/src/lib.rs)

```
...
#[link(wasm_import_module = "wasi_snapshot_preview1")]
extern "C" {
    pub fn sock_accept(fd: Fd, fdflags: u16, new_fd: *mut Fd) -> Errno;
    pub fn sock_send(
        fd: Fd,
        ciovs: *const Ciovec,
        ciovs_len: Size,
        si_flags: u16,
        written: *mut Size,
    ) -> Errno;
    pub fn sock_shutdown(fd: Fd, how: u16) -> Errno;
    pub fn fd_write(fd: Fd, ciovs: *const Ciovec, ciovs_len: Size, nwritten: *mut
Size) -> Errno;
    pub fn proc_exit(rval: Exitcode);
}

const ERRNO_AGAIN: Errno = 6;
const HTTP_RESPONSE: &str =
    "HTTP/1.1 200 OK\r\nContent-Type: text/plain\r\nContent-Length:
18\r\n\r\nHello, WASI 0.1!\r\n";

#[no_mangle]
pub fn _start() {
    let mut fd = 0;
    while let ERRNO_AGAIN = unsafe { sock_accept(3, 0, &mut fd) } {}

    let ciovs = [Ciovec {
        buf: HTTP_RESPONSE.as_ptr(),
```

54 | 第2章 WASI 0.1

```
        buf_len: HTTP_RESPONSE.len(),
    }];
    unsafe { sock_send(fd, ciovs.as_ptr(), ciovs.len(), 0, &mut 0) };


    unsafe { sock_shutdown(fd, 3) };


    unsafe { proc_exit(0) };
}
```

sock_acceptで用いるファイルディスクリプターを「3」に設定していますが、これはWasmtimeから実際に「3」をソケットのファイルディスクリプターとして渡されるためです。

sock_acceptが接続待ちをしている間は「errno = 6 (again)」が返るため、「errno = 6」の間はsock_acceptを繰り返し実行するコードを書く必要があります。

HTTPサーバーにリクエストが届くとsock_acceptが成功し、新たなファイルディスクリプターが返ります。このファイルディスクリプターに対してsock_sendでHTTPレスポンスを返し、sock_shutdownで接続を閉じます。

このプログラムを実際にWasmtimeで実行するには、-S tcplisten=localhost:8080オプションを指定する必要があります。また、最新のWasmtimeはWASI 0.2であることを期待しているため、WASI 0.2ではないことを伝えるための-S preview2=nオプションもあわせて指定しましょう。

Cargo.tomlにlib.crate-type = "cdylib"を追加してWasmにビルドし、実行してみましょう。

```
$ cargo build --release --target wasm32-unknown-unknown
   Compiling socket v0.1.0 (/wasi-book-example/第2章/socket)
    Finished `release` profile [optimized] target(s) in 0.19s
```

```
$ wasmtime run -S tcplisten=localhost:8080 -S preview2=n \
    target/wasm32-unknown-unknown/release/socket.wasm
```

実行すると、プログラムは終了せずに実行状態になります。この状態で別のターミナルを開き、curlコマンドを実行すると、HTTPレスポンスが返ってくることが確認できます。

```
$ curl http://localhost:8080
Hello, WASI 0.1!
```

2.10　WASI 0.1の課題

ここまで、WASI 0.1の主要なインターフェースを解説しました。WASI 0.1はPOSIXを参考にしているとあるように、C言語ライクなポインターを多用したインターフェースであることがわかります。

C言語ライクなインターフェースであることも含めて、WASI 0.1には4つの主な課題があります。

・C言語ライクなインターフェース
・ソケットのサポートが不十分
・仮想化が不十分
・異なる言語で実装されたモジュールを組み合わせることが難しい

C言語ライクなインターフェース

WASI 0.1は、ポインターを多用したインターフェースとなっています。

たとえば`fd_write`関数の仕様上の定義は`fd_write(fd: fd, iovs: ciovec_array) -> Result<size, errno>`ですが、実際のWasm上の実装は`(func (param i32 i32 i32 i32)(result i32))`のようになります。これをRustで表現すると、`fd_write(fd: Fd, ciovs: *const Ciovec, ciovs_len: Size, nwritten: *mut Size) -> Errno`のようになります。

これはWASI 0.1が参考にしたPOSIXのインターフェースが基本的にポインター(もしくは数値)を直接操作するようなインターフェースであること、Wasmのコアスペックにリストや文字列といった高レベルなデータ構造が定義されていないことに起因しています。

WASIはRustをはじめ多くの言語のコンパイルターゲットとしてサポートされることを目指しているため、より抽象度が高く扱いやすいインターフェースを求めています。

ソケットのサポートが不十分

WASI 0.1では既存のソケットに対する操作をサポートしていますが、新たにソケットを作成することをサポートしていません。そのため、WASI 0.1でHTTPサーバーを実装することはできるものの、ソケットに使うファイルディスクリプターを事前に知る必要がある(Wasmtimeの場合は「3」に固定されている)といった制約があります。

Goはバージョン1.21からWASI 0.1をサポートしているため、WASI 0.1を用いてHTTPサーバーを実装することはできます。しかし、ソケットを新たに作成できないという制約を理由に、[7]に標準ライブラリー(net/http)でのHTTPサーバーの実装をサポートしていません。

仮想化が不十分

POSIX由来のインターフェースは、「リソースの対象が何か?」ということを基本的に意識しません。標準出力への書き込みには`fd_write`関数を使い、ファイルへの書き込みにも`fd_write`関数を使います。このとき、標準出力もファイルもファイルディスクリプターを使います。ソケットの場

7.WASI support in Go :https://go.dev/blog/wasi

56　第2章　WASI 0.1

合は使用するインターフェースが異なりますが、これもファイルディスクリプターを使うという点では同じです。

図2.5: WASI 0.1のインターフェース

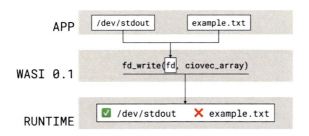

WASIは機能単位に機能を提供、制限することを目標としていますが、「標準出力への書き込みは許可するが、ファイルへの書き込みは許可しない」という制限を行いたい場合、WASI 0.1の仕様ではなく、`fd_write`関数をランタイムがどう実装しているかに依存してしまいます。そのため、WASI 0.1の仕様では、機能単位の仮想化は不十分であるという結論になりました。

異なる言語で実装されたモジュールを組み合わせることが難しい

言語が異なれば、その言語が扱うメモリー上のデータ構造も異なります。Rustで扱うデータとGoで扱うデータが同じようなデータ型であっても、メモリー上は異なるデータ構造ということがあります。そのため、RustからビルドされたWasmモジュールとGoからビルドされたWasmモジュール間でデータの受け渡しを行う場合、メモリー上のデータ構造も含めてデータ構造を決定する必要がありますが、WASI 0.1にはそういったデータ構造を決めるための仕様はありません。

図2.6: 異なる言語で実装されたモジュールを組み合わせるには?

そのため、WASI 0.1の仕様では異なる言語で実装されたWasmを簡単に組み合わせて使うことができません。

2.11 まとめ

WASI 0.1に定義されている主要なインターフェースについて解説しました。WASI 0.1のインターフェースを用いた実装例では、Rustで必要なデータ構造などを定義しましたが、Rustのwasiクレート[8](バージョン0.11.0)を使うと、より簡単に実装することができます。

8.https://crates.io/crates/wasi

リスト2.29: wasi クレートを使用した実装例(wasi-crate/src/main.rs)

```rust
use wasi::{fd_write, Ciovec, FD_STDOUT};

#[export_name = "_start"]
fn main() {
    let message = b"Hello, world!\n";
    let ciovec = [Ciovec {
        buf: message.as_ptr(),
        buf_len: message.len(),
    }];

    unsafe { fd_write(FD_STDOUT, &ciovec).unwrap() };
}
```

　また、WASI 0.1には主に4つの課題がありました。これらの課題を解決することを踏まえて、WASI 0.2の仕様が策定されています。

第3章 コンポーネントモデル

3.1 コンポーネントモデル

コンポーネントモデル[1]はWasmの拡張仕様で、異なる言語からビルドされたWasmを組み合わせて利用するためのABI(Application Binary Interface)とインターフェースを定義するためのIDL(Interface Description Language)を提供します。ウェブブラウザー上で広くサポートされているWasmの仕様はコアスペック[2]と呼ばれるもので、Wasmの基本的な機能を提供しています。WASI 0.2は、このコンポーネントモデルをベースに仕様が策定されました。

コアスペックがモジュールという単位でWasmを扱うのに対して、コンポーネントモデルはモジュールを内包したコンポーネントという単位でWasmを扱います。そのため、モジュールとコンポーネントはバイナリーフォーマットも異なるものとなっており、2024年7月時点ではウェブブラウザーなどでコンポーネントモデルはサポートされていません。

図3.1: コアスペックとコンポーネントモデルのバイナリーフォーマットの違い

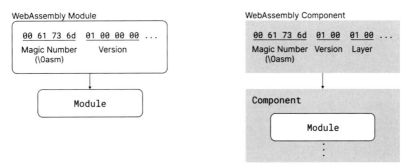

3.2 コンポーネントモデルの生まれた背景

WasmはCPUやOSに依存しないバイナリーフォーマットであるため、RustやGoなどの言語からWasmにビルドすることでWasmランタイムがあれば、どのような環境でも実行することができます。しかし、Wasmのコアスペックで定義しているのは特定のアーキテクチャーに依存しないバイナリーフォーマットの仕様であり、メモリー上のデータの扱いについては定義していません。

言語が異なれば同じようなデータ構造であっても、メモリー上のデータ構造は異なります。そのため、異なる言語からビルドされたWasm間でデータのやり取りを行うためには、メモリー上のデータ構造も含めてデータ構造を決める必要があります。

---

1. コンポーネントモデル :https://component-model.bytecodealliance.org
2. コアスペック :https://webassembly.github.io/spec/core/

コンポーネントモデルはABI(Application Binary Interface)を定義することで、どの言語でも共通のデータ構造を扱えるようにします。これにより、異なる言語からビルドされたWasmであっても共通のデータ構造を扱えるため、どの言語からビルドされたWasmかを意識することなく組み合わせて実行することができるようになります。

3.3　WIT

ABIを決定するだけでは、C言語ライクなインターフェースの操作を行うことからは脱却することはできません。コンポーネントモデルではインターフェースの仕様を記述するためのIDL(Interface Description Language)としてWIT(WebAssembly Interface Type)を提供します。IDLとはその名の通り、インターフェースの仕様を記述するための言語です。コンポーネントモデルではWITを用いてインターフェースの仕様を記述し、そのインターフェースを満たすコードを書くことで、WITに対応するABIを持つコンポーネントにビルドすることができます。

そのため、異なる言語からビルドされたWasmを組み合わせて実行する必要がある場合、共通のWIT定義を用いることで、簡単にコンポーネントを組み合わせることができるようになります。

WITの例

WITは次のように、example:helloパッケージとhelloインターフェースを定義することができます。

リスト3.1: example:hello パッケージ (example/hello.wit)

```
package example:hello@0.1.0;

interface hello {
    greet: func(name: string) -> string;
}

world greeter {
    export hello;
}
```

このWIT定義は、次のように実装することができます[3]。

3.ComponentizeJS:https://github.com/bytecodealliance/ComponentizeJS

60　第3章　コンポーネントモデル

リスト3.2: JavaScriptを用いたWIT実装例 (example/hello.js)

```javascript
export const hello = {
  greet(name) {
    return `Hello, ${name}!`;
  },
};
```

```
$ jco componentize hello.js --wit hello.wit -o hello.wasm
OK Successfully written hello.wasm.
```

また、このexample:helloパッケージのhelloインターフェースを利用する別の
example:componentパッケージを定義して、別の言語で実装することもできます。

リスト3.3: example:component パッケージ (example/component/wit/world.wit)

```
package example:component@0.1.0;

world hello {
    import example:hello/hello@0.1.0;
}
```

リスト3.4: Rustを用いたWIT実装例 (example/component/src/main.rs)

```rust
#[allow(warnings)]
mod bindings;

use bindings::example::hello::hello::greet;

fn main() {
    let greet = greet("asuka");
    println!("{greet}");
}
```

```
$ cargo component build --release
...
```

これらふたつのコンポーネントを組み合わせることで、実行することができます。

第3章 コンポーネントモデル　61

```
$ wasm-tools compose -o compose.wasm -d hello.wasm component.wasm
...
```

```
$ wasmtime run compose.wasm
Hello, asuka!
```

このように、WITはWAT[4]にはなかった文字列(string)などの抽象度の高い型を扱えるようになっています。

パッケージ名

WITはひとつのディレクトリーに複数のWITファイルを記述することができますが、ひとつのディレクトリーにはひとつのパッケージしか定義することはできません。パッケージ名はネームスペース、パッケージ、バージョンの3つの要素で構成されます。

リスト3.5: パッケージ名の例
```
package example:hello@0.1.0;
```

この例の場合、ネームスペースがexample、パッケージ名がhello、バージョンが0.1.0となります。このうちネームスペースとバージョンは省略可能で、example:hello(バージョンを省略)やhello@0.1.0(ネームスペースを省略)のように記述することもできます。また、バージョンにはセマンティックバージョニング[5]を採用しています。

インターフェース

インターフェースはWITの中核となる概念で、関数と型の集合を定義できます。

リスト3.6: インターフェースの例
```
interface hello {
    greet: func(name: string) -> string;
}
```

この例では、helloインターフェースにgreet関数を定義しています。インターフェースはあくまで関数と型の集合を定義するだけのもので、実際の実装はWITファイルには含まれていません。

ワールド

ワールドはWITパッケージにおける実装単位で、Wasmコンポーネントに実装するインターフェー

4.WAT : WebAssembly Text Format

5. セマンティックバージョニング :https://semver.org/lang/ja

62　　第3章　コンポーネントモデル

スの集合を定義します。

リスト3.7: ワールドの例

```
world greeter {
    export hello;
}
```

この例では、helloインターフェースを実装するgreeterワールドを定義しています。このgreeter
ワールドはhelloインターフェースをエクスポートしているため、このワールドを実装したコンポー
ネントのhelloインターフェースを使用することができます。

また、example:helloパッケージのhelloインターフェースを使用する場合は、インポート構文
を用います。次のように書くことで、example:helloパッケージのhelloインターフェースを使用
することができます。

リスト3.8: インポートの例

```
world hello {
    import example:hello/hello@0.1.0;
}
```

3.4　example:helloパッケージの実装

先ほどのexample:helloパッケージをRustで実装してみましょう。Rustのプロジェクトを作成
し、WITファイルを作成します。

```
$ cargo component new --lib hello
...
```

リスト3.9: hello/wit/hello.wit

```
package example:hello@0.1.0;

interface hello {
    greet: func(name: string) -> string;
}

world greeter {
    export hello;
}
```

Cargo.tomlのpackage.metadata.component.packageを、作成したWITファイルのパッケージ名に合わせて書き換えましょう。

リスト3.10: hello/Cargo.toml

```
...

[package.metadata.component]
package = "example:hello"


...
```

この状態でビルドすると、WIT定義の実装が正しく行われていないためにエラーとなりますが、WIT定義をもとにsrc/bindings.rsが自動生成されます。

```
$ cargo component build
  Generating bindings for hello (src/bindings.rs)
    Compiling wit-bindgen-rt v0.27.0
    Compiling bitflags v2.6.0
    Compiling hello v0.1.0 (/wasi-book-example/第3章/hello)
error[E0432]: unresolved import `bindings::Guest`
 --> src/lib.rs:4:5
   |
4 | use bindings::Guest;
   |     ^^^^^^^^^^^^^^^ no `Guest` in `bindings`
   |
help: consider importing this trait instead
   |
4 | use crate::bindings::exports::example::hello::hello::Guest;
   |     ~~~~~~~~~~~~~~~~~~~~~~~~~~~~~~~~~~~~~~~~~~~~~~~~~~~~~~~

For more information about this error, try `rustc --explain E0432`.
error: could not compile `hello` (lib) due to 1 previous error
```

このsrc/bindings.rsを利用して、src/lib.rsを次のように書き換えます。

リスト3.11: hello/src/lib.rs

```
#[allow(warnings)]
mod bindings;


use crate::bindings::exports::example::hello::hello::Guest;


struct Component;


impl Guest for Component {
    fn greet(name: String) -> String {
```

64　第3章　コンポーネントモデル

```
        format!("Hello, {name}!")
    }
}

bindings::export!(Component with_types_in bindings);
```

ビルドすると、example:helloパッケージを実装したWasmコンポーネントができます。

```
$ cargo component build --release
   Compiling wit-bindgen-rt v0.27.0
   Compiling bitflags v2.6.0
   Compiling hello v0.1.0 (/wasi-book-example/第3章/hello)
    Finished `release` profile [optimized] target(s) in 0.57s
    Creating component target/wasm32-wasi/release/hello.wasm
```

3.5 example:componentパッケージの実装

次に、example:helloを利用するWasmコンポーネントを実装しましょう。Rustのプロジェクト
を作成し、WITファイルを作成します。

```
$ cargo component new component
...
```

リスト3.12: component/wit/world.wit
```
package example:component@0.1.0;

world hello {
    import example:hello/hello@0.1.0;
}
```

Cargo.tomlのpackage.metadata.component.packageをWITファイルのパッケージ名に合わせて
書き換え、ビルドすると、次のようなエラーが出ます。

```
$ cargo component build
error: failed to create a target world for package `component` ...

Caused by:
    0: failed to merge local target `/wasi-book-example/第3章/component/wit`
    1: package not found
```

第3章 コンポーネントモデル | 65

```
        --> /wasi-book-example/第3章/component/wit/world.wit:4:12
          |
      4 |     import example:hello/hello@0.1.0;
          |            ^------------
...
```

example:helloパッケージのWIT定義はcomponent/witではなくhello/witにありますが、コンパイラーはそのことを知りません。このビルドを通すには、example:helloパッケージのWIT定義がどこにあるかコンパイラーに伝える必要があります。

次のコマンドを実行することで、Cargo.tomlに依存関係の情報が追加されます。

```
$ cargo component add --path ../hello/wit --target example:hello
      Added dependency `example:hello` from path `../hello/wit`
```

リスト3.13: component/Cargo.toml

```
...

[package.metadata.component.target.dependencies]
"example:hello" = { path = "../hello/wit" }

...
```

Cargo.tomlを開くと、`package.metadata.component.target.dependencies`にexample:helloのWIT定義のパスが追加されていることがわかります。この状態でビルドすると、src/bindings.rsが自動生成されます。

```
$ cargo component build
  Generating bindings for component (src/bindings.rs)
   Compiling wit-bindgen-rt v0.27.0
   Compiling bitflags v2.6.0
   Compiling component v0.1.0 (/wasi-book-example/第3章/component)
    Finished `dev` profile [unoptimized + debuginfo] target(s) in 0.55s
    Creating component target/wasm32-wasi/debug/component.wasm
```

example:helloの利用

src/bindings.rsが生成されたのを確認し、src/main.rsを次のように書き換えビルドすると、component.wasmが作成されます。

66 | 第3章 コンポーネントモデル

リスト3.14: component/src/main.rs

```
#[allow(warnings)]
mod bindings;

use bindings::example::hello::hello::greet;

fn main() {
    let greet = greet("asuka");
    println!("{greet}");
}
```

```
$ cargo component build --release
   Compiling wit-bindgen-rt v0.27.0
   Compiling bitflags v2.6.0
   Compiling component v0.1.0 (/wasi-book-example/第3章/component)
    Finished `release` profile [optimized] target(s) in 0.65s
    Creating component target/wasm32-wasi/release/component.wasm
```

ビルドしたcomponent.wasmを実行すると、インポートしているexample:hello/hello@0.1.0の実装が見つからない旨のエラーが出ます。

```
$ wasmtime run target/wasm32-wasi/release/component.wasm
Error: failed to run main module `target/wasm32-wasi/release/component.wasm`

Caused by:
    0: component imports instance `example:hello/hello@0.1.0`, but a matching
implementation was not found in the linker
    1: instance export `greet` has the wrong type
    2: function implementation is missing
```

このエラーを解決するには、example:helloパッケージの実装であるhello.wasmとexample:componentパッケージの実装であるcomponent.wasmをひとつのコンポーネントにまとめる必要があります。次のようにwasm-toolsコマンドを使うことで、hello.wasmとcomponent.wasmをひとつのコンポーネントにまとめることができます。

```
$ wasm-tools compose -o compose.wasm \
    -d hello/target/wasm32-wasi/release/hello.wasm \
    component/target/wasm32-wasi/release/component.wasm
WARNING: `wasm-tools compose` has been deprecated.

Please use `wac` instead. You can find more information about `wac` at
https://github.com/bytecodealliance/wac.
...
```

第3章　コンポーネントモデル　　67

```
composed component `compose.wasm`
```

wasm-tools composeコマンドを実行するとcompose.wasmを作成することができますが、wasm-tools composeコマンドの代わりにwacコマンドの使用を促すメッセージが表示されます。

WACについては5章であらためて紹介しますが、wacコマンドを用いてビルドする場合は次のようになります。

```
$ wac plug component/target/wasm32-wasi/release/component.wasm \
    --plug hello/target/wasm32-wasi/release/hello.wasm \
    -o compose.wasm
```

このcompose.wasmを実行すると、期待する実行結果を得ることができます。

```
$ wasmtime run compose.wasm
Hello, asuka!
```

3.6 まとめ

この章ではコンポーネントモデルの基本的な概念とWITファイルの書き方、コンポーネントの実装方法を紹介しました。コンポーネントモデルの要点は、次のふたつです。

1．より抽象度の高いインターフェース定義を提供することで、容易にコンポーネントを組み合わせることができる。

2．ABIを通して異なる言語からビルドされたコンポーネントも組み合わせることができる。

コンポーネントモデルを使用することで、他者にWasmコンポーネントを提供しやすくなるだけでなく、他者のWasmコンポーネントを利用することも容易になります。今後、コンポーネントモデルがWasmを使用する際の基盤になることは間違いないでしょう。

68　　第3章　コンポーネントモデル

第4章 WASI 0.2

4.1 WASI 0.2(プレビュー2)

　WASI 0.2(プレビュー2)ではインターフェースの大幅な見直しが行われ、前章で紹介したコンポーネントモデルをベースに、インターフェースの再設計が行われました。WASI 0.1では、ひとつの `wasi_snapshot_preview1` モジュールに全てのインターフェースがまとめられていましたが、WASI 0.2では機能ごとに別々のコンポーネントに分けられています。

表4.1: WASI 0.2のパッケージ一覧

WITパッケージ名	バージョン	URL
wasi:cli	0.2.0	https://github.com/WebAssembly/wasi-cli
wasi:io	0.2.0	https://github.com/WebAssembly/wasi-io
wasi:clocks	0.2.0	https://github.com/WebAssembly/wasi-clocks
wasi:random	0.2.0	https://github.com/WebAssembly/wasi-random
wasi:filesystem	0.2.0	https://github.com/WebAssembly/wasi-filesystem
wasi:sockets	0.2.0	https://github.com/WebAssembly/wasi-sockets
wasi:http	0.2.0	https://github.com/WebAssembly/wasi-http

　前章の最後に作成したcompose.wasmのWIT定義を `wasm-tools` コマンドを使用して表示すると、次のようになります。

```
$ wasm-tools component wit compose.wasm
package root:component;

world root {
  import wasi:cli/environment@0.2.0;
  import wasi:cli/exit@0.2.0;
  import wasi:io/error@0.2.0;
  import wasi:io/streams@0.2.0;
  import wasi:cli/stdin@0.2.0;
  import wasi:cli/stdout@0.2.0;
  import wasi:cli/stderr@0.2.0;
  import wasi:clocks/wall-clock@0.2.0;
  import wasi:filesystem/types@0.2.0;
  import wasi:filesystem/preopens@0.2.0;

  export wasi:cli/run@0.2.0;
}
```

このインポート/エクスポートされているwasiネームスペースで始まるインターフェースが、WASI
0.2以降のインターフェースになります。

4.2 wasi:cliパッケージ

wasi:cliパッケージには、環境変数や引数を取得するためのenvironmentインターフェース、
標準入力を扱うためのstdinインターフェース、標準出力を扱うためのstdoutインターフェース、
Wasmを実行するためのrunインターフェースなどが定義されています。

リスト4.1: wasi:cli/environmentインターフェース

```
interface environment {
    get-environment: func() -> list<tuple<string, string>>;
    get-arguments: func() -> list<string>;
    initial-cwd: func() -> option<string>;
}
```

リスト4.2: wasi:cli/stdinインターフェース

```
interface stdin {
    use wasi:io/streams@0.2.0.{input-stream};

    get-stdin: func() -> input-stream;
}
```

リスト4.3: wasi:cli/stdoutインターフェース

```
interface stdout {
    use wasi:io/streams@0.2.0.{output-stream};

    get-stdout: func() -> output-stream;
}
```

リスト4.4: wasi:cli/run インターフェース

```
interface run {
    run: func() -> result;
}
```

WASI 0.1では_start関数がエントリーポイントとなっていましたが、WASI 0.2ではrunインター
フェースがプログラムのエントリーポイントとなります。

実行

Rustのmain関数を用いずに、wasi:cli/runインターフェースを実装してWasmを実行してみま
しょう。Rustのプロジェクトを作成し、WITファイルを用意します。

```
$ cargo component new --lib cli_run
...
```

リスト4.5: cli_run/wit/world.wit

```
package component:cli-run;

world example {
    export wasi:cli/run@0.2.0;
}
```

wasi:cli/runインターフェースの定義を定義するwit/cli/run.witを用意し、依存関係に追加します。

リスト4.6: cli_run/wit/cli/run.wit

```
package wasi:cli@0.2.0;

interface run {
    run: func() -> result;
}
```

```
$ cargo component add --path ./wit/cli --target wasi:cli
      Added dependency `wasi:cli` from path `./wit/cli`
```

これで、WITファイルの準備が完了しました。ビルドするとエラーが出ますが、WIT定義から
src/bindings.rsが生成されます。

第4章　WASI 0.2 ｜ 71

```
$ cargo component build
  Generating bindings for cli_run (src/bindings.rs)
   Compiling wit-bindgen-rt v0.27.0
   Compiling bitflags v2.6.0
   Compiling cli_run v0.1.0 (/wasi-book-example/第4章/cli_run)
error[E0432]: unresolved import `bindings::Guest`
 --> src/lib.rs:4:5
  |
4 | use bindings::Guest;
  |     ^^^^^^^^^^^^^^^ no `Guest` in `bindings`
  |
help: consider importing this trait instead
  |
4 | use crate::bindings::exports::wasi::cli::run::Guest;
  |     ~~~~~~~~~~~~~~~~~~~~~~~~~~~~~~~~~~~~~~~~~~~~~~~~

For more information about this error, try `rustc --explain E0432`.
error: could not compile `cli_run` (lib) due to 1 previous error
```

src/lib.rsを次のように変更しビルドすると、cli_run.wasmが生成されます。

リスト4.7: cli_run/src/lib.rs

```
#[allow(warnings)]
mod bindings;

use bindings::exports::wasi::cli::run::Guest;

struct Component;

impl Guest for Component {
    fn run() -> Result<(), ()> {
        Ok(())
    }
}

bindings::export!(Component with_types_in bindings);
```

```
$ cargo component build --release
   Compiling wit-bindgen-rt v0.27.0
   Compiling bitflags v2.6.0
   Compiling cli_run v0.1.0 (/wasi-book-example/第4章/cli_run)
    Finished `release` profile [optimized] target(s) in 0.69s
    Creating component target/wasm32-wasi/release/cli_run.wasm
```

ビルドされたcli_run.wasmを実行してみましょう。何も表示されなければ、正常に実行されています。

```
$ wasmtime run target/wasm32-wasi/release/cli_run.wasm
```

終了コードを確認すると、0 (正常終了)であることが確認できます。

```
$ echo $?
0
```

run関数の戻り値をErr(())に変更して、ビルドして実行してみましょう。

リスト4.8: cli_run/src/lib.rs (戻り値を Err(())に変更)

```
#[allow(warnings)]
mod bindings;

use bindings::exports::wasi::cli::run::Guest;

struct Component;

impl Guest for Component {
    fn run() -> Result<(), ()> {
        Err(())
    }
}

bindings::export!(Component with_types_in bindings);
```

```
$ cargo component build --release
...
```

```
$ wasmtime run target/wasm32-wasi/release/cli_run.wasm
```

```
$ echo $?
1
```

第4章　WASI 0.2 | 73

終了コードが1(エラー終了)になっていることが確認できます。また、ビルドしたcli_run.wasmの
WIT定義を確認すると、wasi:cli/runインターフェースがエクスポートされていることが確認で
きます。

```
$ wasm-tools component wit target/wasm32-wasi/release/cli_run.wasm
package root:component;

world root {
    export wasi:cli/run@0.2.0;
}
```

標準出力

wasi:cli/stdoutインターフェースを用いて、標準出力に「Hello, world!」を表示するWasmを
作成してみましょう。Rustのパッケージを作成し、WITファイルを用意します。

```
$ cargo component new --lib cli_stdout
...
```

リスト4.9: cli_stdout/wit/world.wit
```
package component:cli-stdout;

world example {
    import wasi:cli/stdout@0.2.0;
    export wasi:cli/run@0.2.0;
}
```

wit/world.witには、wasi:cli/stdoutインターフェースをインポートし、wasi:cli/runインター
フェースをエクスポートするワールドを定義しています。これらの必要なWIT定義を用意し、依存
関係に追加します。

リスト4.10: cli_stdout/wit/cli/run.wit
```
package wasi:cli@0.2.0;

interface run {
    run: func() -> result;
}
```

74 │ 第4章 WASI 0.2

リスト 4.11: cli_stdout/wit/cli/stdio.wit

```
package wasi:cli@0.2.0;

interface stdout {
    use wasi:io/streams@0.2.0.{output-stream};
    get-stdout: func() -> output-stream;
}
```

```
$ cargo component add --path ./wit/cli --target wasi:cli
        Added dependency `wasi:cli` from path `./wit/cli`
```

wasi:cli/stdout インターフェースはwasi:io/streams インターフェースに依存しているため、wasi:io パッケージの WIT 定義も用意します。

リスト 4.12: cli_stdout/wit/io/streams.wit (一部)

```
package wasi:io@0.2.0;

interface streams {
    use error.{error};
    use poll.{pollable};

...

    resource output-stream {
        check-write: func() -> result<u64, stream-error>;
        write: func(contents: list<u8>) -> result<_, stream-error>;
        blocking-write-and-flush: func(
            contents: list<u8>,
        ) -> result<_, stream-error>;
        flush: func() -> result<_, stream-error>;
        blocking-flush: func() -> result<_, stream-error>;
        subscribe: func() -> pollable;
        write-zeroes: func(len: u64) -> result<_, stream-error>;
        blocking-write-zeroes-and-flush: func(
            len: u64,
        ) -> result<_, stream-error>;
        splice: func(
            src: borrow<input-stream>,
            len: u64,
        ) -> result<u64, stream-error>;
        blocking-splice: func(
```

第4章　WASI 0.2 | 75

```
        src: borrow<input-stream>,
        len: u64,
    ) -> result<u64, stream-error>;
    }
}
```

リスト4.13: cli_stdout/wit/io/error.wit

```
package wasi:io@0.2.0;


interface error {
    resource error {
        to-debug-string: func() -> string;
    }
}
```

リスト4.14: cli_stdout/wit/io/poll.wit

```
package wasi:io@0.2.0;

interface poll {
    resource pollable {
      ready: func() -> bool;
      block: func();
    }

    poll: func(in: list<borrow<pollable>>) -> list<u32>;
}
```

```
$ cargo component add --path ./wit/io --target wasi:io
      Added dependency `wasi:io` from path `./wit/io`
```

この状態でビルドするとエラーが出ますが、WIT定義からsrc/bindings.rsが生成されます。

```
$ cargo component build
  Generating bindings for cli_stdout (src/bindings.rs)
    Compiling wit-bindgen-rt v0.27.0
    Compiling bitflags v2.6.0
    Compiling cli_stdout v0.1.0 (/wasi-book-example/第4章/cli_stdout)
```

76 | 第4章 WASI 0.2

```
...
```

src/lib.rsを次のように変更し、ビルドすると、cli_stdout.wasmが生成されます。

リスト4.15: cli_stdout/src/lib.rs

```rust
#[allow(warnings)]
mod bindings;

use bindings::exports::wasi::cli::run::Guest;
use bindings::wasi::cli::stdout::get_stdout;

struct Component;
impl Guest for Component {
    fn run() -> Result<(), ()> {
        let stdout = get_stdout();
        stdout.blocking_write_and_flush(b"Hello, world!\n").unwrap();
        Ok(())
    }
}

bindings::export!(Component with_types_in bindings);
```

```
$ cargo component build --release
   Compiling wit-bindgen-rt v0.27.0
   Compiling bitflags v2.6.0
   Compiling cli_stdout v0.1.0 (/wasi-book-example/第4章/cli_stdout)
    Finished `release` profile [optimized] target(s) in 0.63s
    Creating component target/wasm32-wasi/release/cli_stdout.wasm
```

ビルドされたcli_stdout.wasmを実行すると、標準出力に「Hello, world!」が表示されます。

```
$ wasmtime run target/wasm32-wasi/release/cli_stdout.wasm
Hello, world!
```

WASI 0.1と比較すると、WASI 0.1が構造体とポインターを用いたインターフェースだったのに対して、WASI 0.2ではバイト列を直接渡せるインターフェースとなっており、より扱いやすいインターフェースになっていることがわかります。

標準入力

　wasi:cli/stdinインターフェースを用いて、標準入力から文字列を読み込むWasmを作成してみましょう。Rustのパッケージを作成し、WITファイルを用意します。

第4章　WASI 0.2　　77

```
$ cargo component new --lib cli_stdin
...
```

リスト4.16: cli_stdin/wit/world.wit

```
package component:cli-stdout;

world example {
    import wasi:cli/stdin@0.2.0;
    import wasi:cli/stdout@0.2.0;
    export wasi:cli/run@0.2.0;
}
```

wit/world.witは、先ほどの標準出力の例にwasi:cli/stdinインターフェースのインポートを追加したものとなります。標準出力の例同様に必要なWIT定義を用意し、依存関係に追加しましょう。

リスト4.17: cli_stdin/wit/cli/run.wit

```
package wasi:cli@0.2.0;

interface run {
    run: func() -> result;
}
```

リスト4.18: cli_stdin/wit/cli/stdio.wit

```
package wasi:cli@0.2.0;

interface stdin {
    use wasi:io/streams@0.2.0.{input-stream};
    get-stdin: func() -> input-stream;
}

interface stdout {
    use wasi:io/streams@0.2.0.{output-stream};
    get-stdout: func() -> output-stream;
}
```

78 | 第4章 WASI 0.2

```
$ cargo component add --path ./wit/cli --target wasi:cli
        Added dependency `wasi:cli` from path `./wit/cli`
```

wasi:ioパッケージのWIT定義も同様に用意し、依存関係に追加します。

リスト4.19: cli_stdin/wit/io/streams.wit (一部)

```
package wasi:io@0.2.0;

interface streams {
    use error.{error};
    use poll.{pollable};

...

    resource input-stream {
        read: func(len: u64) -> result<list<u8>, stream-error>;
        blocking-read: func(len: u64) -> result<list<u8>, stream-error>;
        skip: func(len: u64) -> result<u64, stream-error>;
        blocking-skip: func(len: u64) -> result<u64, stream-error>;
        subscribe: func() -> pollable;
    }

...
}
```

リスト4.20: cli_stdin/wit/io/error.wit

```
package wasi:io@0.2.0;

interface error {
    resource error {
        to-debug-string: func() -> string;
    }
}
```

リスト4.21: cli_stdin/wit/io/poll.wit

第4章 WASI 0.2 | 79

```
package wasi:io@0.2.0;

interface poll {
    resource pollable {
        ready: func() -> bool;
        block: func();
    }


    poll: func(in: list<borrow<pollable>>) -> list<u32>;
}
```

```
$ cargo component add --path ./wit/io --target wasi:io
        Added dependency `wasi:io` from path `./wit/io`
```

src/lib.rsを次のように変更し、ビルドするとcli_stdin.wasmが生成されます。

リスト4.22: cli_stdin/src/lib.rs

```
#[allow(warnings)]
mod bindings;

use bindings::exports::wasi::cli::run::Guest;
use bindings::wasi::cli::stdin::get_stdin;
use bindings::wasi::cli::stdout::get_stdout;

struct Component;
impl Guest for Component {
    fn run() -> Result<(), ()> {
        let stdout = get_stdout();
        stdout.blocking_write_and_flush(b"Your name: ").unwrap();

        let stdin = get_stdin();
        let name = stdin.blocking_read(1024).unwrap();
        let name = String::from_utf8(name).unwrap();
        let name = name.trim();

        stdout
            .blocking_write_and_flush(format!("Hello, {name}!\n").as_bytes())
            .unwrap();
        Ok(())
    }
}
```

80 | 第4章 WASI 0.2

```
bindings::export!(Component with_types_in bindings);
```

```
$ cargo component build --release
  Generating bindings for cli_stdin (src/bindings.rs)
   Compiling wit-bindgen-rt v0.27.0
   Compiling bitflags v2.6.0
   Compiling cli_stdin v0.1.0 (/wasi-book-example/第4章/cli_stdin)
    Finished `release` profile [optimized] target(s) in 0.72s
    Creating component target/wasm32-wasi/release/cli_stdin.wasm
```

ビルドされたcli_stdin.wasmを実行すると、標準入力から入力した文字列が表示されます。

```
$ wasmtime run target/wasm32-wasi/release/cli_stdin.wasm
Your name: asuka
Hello, asuka!
```

環境変数

環境変数を取得するには、environmentインターフェースのget-environment関数を使用します。Rustのパッケージを作成し、WITファイルを用意します。

```
$ cargo component new --lib cli_env
...
```

リスト4.23: cli_env/wit/world.wit

```
package component:cli-env;

world example {
    import wasi:cli/environment@0.2.0;
    import wasi:cli/stdout@0.2.0;
    export wasi:cli/run@0.2.0;
}
```

必要なWIT定義を用意し、依存関係に追加します。

リスト4.24: cli_env/wit/cli/run.wit

第4章　WASI 0.2　　81

```
package wasi:cli@0.2.0;

interface run {
    run: func() -> result;
}
```

リスト 4.25: cli_env/wit/cli/environment.wit

```
package wasi:cli@0.2.0;

interface environment {
  get-environment: func() -> list<tuple<string, string>>;
  get-arguments: func() -> list<string>;
  initial-cwd: func() -> option<string>;
}
```

```
$ cargo component add --path ./wit/cli --target wasi:cli
        Added dependency `wasi:cli` from path `./wit/cli`
```

wasi:ioパッケージのWIT定義も用意し、依存関係に追加します。

リスト 4.26: cli_env/wit/io/streams.wit

```
package wasi:io@0.2.0;

interface streams {
    use error.{error};
    use poll.{pollable};

...

    resource output-stream {
...

        blocking-write-and-flush: func(
            contents: list<u8>,
        ) -> result<_, stream-error>;

...
}
```

82 │ 第4章 WASI 0.2

リスト 4.27: cli_env/wit/io/error.wit

```
package wasi:io@0.2.0;


interface error {
    resource error {
        to-debug-string: func() -> string;
    }
}
```

リスト 4.28: cli_env/wit/io/poll.wit

```
package wasi:io@0.2.0;


interface poll {
    resource pollable {
      ready: func() -> bool;
      block: func();
    }

    poll: func(in: list<borrow<pollable>>) -> list<u32>;
}
```

```
$ cargo component add --path ./wit/io --target wasi:io
        Added dependency `wasi:io` from path `./wit/io`
```

src/lib.rsを次のように変更し、ビルドするとcli_env.wasmが生成されます。

リスト 4.29: cli_env/src/lib.rs

```
#[allow(warnings)]
mod bindings;

use bindings::exports::wasi::cli::run::Guest;
use bindings::wasi::cli::environment::get_environment;
use bindings::wasi::cli::stdout::get_stdout;


struct Component;


impl Guest for Component {
    fn run() -> Result<(), ()> {
        let env = get_environment();
```

第4章　WASI 0.2　83

```
        let stdout = get_stdout();
        for (name, value) in env {
            stdout
                .blocking_write_and_flush(format!("{name}={value}\n").as_bytes())
                .unwrap();
        }
        Ok(())
    }
}

bindings::export!(Component with_types_in bindings);
```

```
$ cargo component build --release
  Generating bindings for cli_env (src/bindings.rs)
   Compiling wit-bindgen-rt v0.27.0
   Compiling bitflags v2.6.0
   Compiling cli_env v0.1.0 (/wasi-book-example/第4章/cli_env)
    Finished `release` profile [optimized] target(s) in 0.73s
    Creating component target/wasm32-wasi/release/cli_env.wasm
```

　Wasmtimeの--envオプションを用いて環境変数を設定して実行することで、Wasmに環境変数を渡すことができます。

```
$ wasmtime run --env FOO=1 --env BAR=2 --env BAZ=3 \
    target/wasm32-wasi/release/cli_env.wasm
FOO=1
BAR=2
BAZ=3
```

　実行環境の全ての環境変数を渡したい場合は、-S inherit-envオプションを指定して実行することもできます。

```
$ wasmtime run -S inherit-env target/wasm32-wasi/release/cli_env.wasm
USER=asuka
...
```

引数

　引数を取得するには、environmentインターフェースのget-arguments関数を使用します。Rustのパッケージを作成し、WITファイルを用意します。

84 ｜ 第4章　WASI 0.2

```
$ cargo component new --lib cli_args
...
```

リスト4.30: cli_args/wit/world.wit

```
package component:cli-args;

world example {
    import wasi:cli/environment@0.2.0;
    import wasi:cli/stdout@0.2.0;
    export wasi:cli/run@0.2.0;
}
```

必要なWIT定義を用意し、依存関係に追加します。

リスト4.31: cli_args/wit/cli/run.wit

```
package wasi:cli@0.2.0;

interface run {
    run: func() -> result;
}
```

リスト4.32: cli_args/wit/cli/environment.wit

```
package wasi:cli@0.2.0;

interface environment {
  get-environment: func() -> list<tuple<string, string>>;
  get-arguments: func() -> list<string>;
  initial-cwd: func() -> option<string>;
}
```

リスト4.33: cli_args/wit/cli/stdio.wit

```
package wasi:cli@0.2.0;

interface stdout {
    use wasi:io/streams@0.2.0.{output-stream};
```

第4章　WASI 0.2　85

```
    get-stdout: func() -> output-stream;
}
```

```
$ cargo component add --path ./wit/cli --target wasi:cli
        Added dependency `wasi:cli` from path `./wit/cli`
```

リスト 4.34: cli_args/wit/io/streams.wit

```
package wasi:io@0.2.0;

interface streams {
    use error.{error};
    use poll.{pollable};

...

    resource output-stream {
...
        blocking-write-and-flush: func(
            contents: list<u8>,
        ) -> result<_, stream-error>;
...
}
```

リスト 4.35: cli_args/wit/io/error.wit

```
package wasi:io@0.2.0;

interface error {
    resource error {
        to-debug-string: func() -> string;
    }
}
```

86 第4章 WASI 0.2

リスト4.36: cli_args/wit/io/poll.wit

```
package wasi:io@0.2.0;

interface poll {
    resource pollable {
      ready: func() -> bool;
      block: func();
    }


    poll: func(in: list<borrow<pollable>>) -> list<u32>;
}
```

```
$ cargo component add --path ./wit/io --target wasi:io
      Added dependency `wasi:io` from path `./wit/io`
```

src/lib.rsを次のように変更し、ビルドすると、cli_args.wasmが生成されます。

リスト4.37: cli_args/src/lib.rs

```
#[allow(warnings)]
mod bindings;

use bindings::exports::wasi::cli::run::Guest;
use bindings::wasi::cli::environment::get_arguments;
use bindings::wasi::cli::stdout::get_stdout;

struct Component;

impl Guest for Component {
    fn run() -> Result<(), ()> {
        let args = get_arguments();
        let stdout = get_stdout();
        for arg in args {
            stdout
                .blocking_write_and_flush(format!("{arg}\n").as_bytes())
                .unwrap();
        }
        Ok(())
    }
}

bindings::export!(Component with_types_in bindings);
```

第4章　WASI 0.2　87

実行すると、引数を取得できているのを確認できます。

```
$ cargo component build --release
  Generating bindings for cli_args (src/bindings.rs)
    Compiling wit-bindgen-rt v0.27.0
    Compiling bitflags v2.6.0
    Compiling cli_args v0.1.0 (/wasi-book-example/第4章/cli_args)
     Finished `release` profile [optimized] target(s) in 0.72s
      Creating component target/wasm32-wasi/release/cli_args.wasm
```

```
$ wasmtime run target/wasm32-wasi/release/cli_args.wasm foo bar
cli_args.wasm
foo
bar
```

commandワールド

ここまでwasi:cliパッケージに定義されているインターフェースを紹介しました。wasi:cliには、インターフェースの他にもcommandワールドが定義されています。

リスト4.38: wasi:cli/command ワールド

```
package wasi:cli@0.2.0;

world command {
    include wasi:clocks/imports@0.2.0;
    include wasi:filesystem/imports@0.2.0;
    include wasi:sockets/imports@0.2.0;
    include wasi:random/imports@0.2.0;
    include wasi:io/imports@0.2.0;

    import environment;
    import exit;
    import stdin;
    import stdout;
    import stderr;
    import terminal-input;
    import terminal-output;
    import terminal-stdin;
    import terminal-stdout;
    import terminal-stderr;
```

```
    export run;
}
```

このcommandワールドは、Wasmを実行するのに必要な汎用的なインターフェースをまとめたものです。Rustのmain関数をWASI 0.2向けにビルドした際に実装されるのは、このwasi:cli/commandワールドです。

Rustの新しいパッケージを用意し、main関数を実装してみましょう。

```
$ cargo component new cli_command
...
```

リスト 4.39: cli_command/src/main.rs

```
#[allow(warnings)]
mod bindings;

fn main() {
    println!("Hello, world!");
}
```

ビルドして実行すると、「Hello, world!」が表示されます。

```
$ cargo component build --release
  Generating bindings for cli_command (src/bindings.rs)
   Compiling wit-bindgen-rt v0.27.0
   Compiling bitflags v2.6.0
   Compiling cli_command v0.1.0 (/wasi-book-example/第4章/cli_command)
    Finished `release` profile [optimized] target(s) in 0.54s
    Creating component target/wasm32-wasi/release/cli_command.wasm
```

```
$ wasmtime run target/wasm32-wasi/release/cli_command.wasm
Hello, world!
```

このときに実装されているのが、wasi:cli/commandワールドになります。

第4章　WASI 0.2　89

```
$ wasm-tools component wit target/wasm32-wasi/release/cli_command.wasm
package root:component;

world root {
  import wasi:cli/environment@0.2.0;
  import wasi:cli/exit@0.2.0;
  import wasi:io/error@0.2.0;
  import wasi:io/streams@0.2.0;
  import wasi:cli/stdin@0.2.0;
  import wasi:cli/stdout@0.2.0;
  import wasi:cli/stderr@0.2.0;
  import wasi:clocks/wall-clock@0.2.0;
  import wasi:filesystem/types@0.2.0;
  import wasi:filesystem/preopens@0.2.0;

  export wasi:cli/run@0.2.0;
}
```

2024年7月時点でRustから`wasi:cli/command`ワールドを実装できるのは`cargo component`コマンドのみですが、2024年末までには`wasm32-wasip2`をビルドターゲットとして、`cargo build`コマンドから`wasi:cli/command`ワールドを実装できるようになるでしょう[1]。

4.3   wasi:ioパッケージ

`wasi:io`パッケージには、ファイルやネットワークなどの入出力を扱うためのインターフェースが定義されています。

streamsインターフェース

streamsインターフェースには、入出力ストリームを扱うための型が定義されています。

リスト4.40: wasi:io/streams インターフェース

```
package wasi:io@0.2.0;

interface streams {
    use error.{error};
    use poll.{pollable};

    variant stream-error {
        last-operation-failed(error),
        closed,
    }
```

1.https://blog.rust-lang.org/2024/04/09/updates-to-rusts-wasi-targets.html

```
    resource input-stream {
...
    }

    resource output-stream {
...
    }
}
```

input-streamとoutput-streamは、それぞれ入力ストリームと出力ストリームを表しています。使用方法については、wasi:cliパッケージの標準入力や標準出力の実装例の通りです。streamsインターフェースは単体で使用する想定ではなく、何かしらのIOストリーム(標準入出力やファイル、ネットワークなど)を扱う場合に使用します。

pollインターフェース

pollインターフェースは、IOストリームの実行を待つためのpollableリソースを定義しています。pollableリソースを使用してIOの実行待ちを行える他、poll関数に複数のpollableリソースを渡して実行を待つことができます。

リスト4.41: wasi:io/poll インターフェース

```
package wasi:io@0.2.0;

interface poll {
    resource pollable {
      ready: func() -> bool;
      block: func();
    }

    poll: func(in: list<borrow<pollable>>) -> list<u32>;
}
```

errorインターフェース

errorインターフェースは、IOエラーを表すためのerrorリソースを定義しています。errorリソース単体で使用することはなく、IOストリームで何らかのエラーが発生した場合のエラーを表現する用途で使用します。

リスト 4.42: wasi:io/error インターフェース

```
package wasi:io@0.2.0;

interface error {
    resource error {
        to-debug-string: func() -> string;
    }
}
```

4.4 wasi:clocks パッケージ

wasi:clocks パッケージは経過時間を取得するための monotonic-clock と、現在の時刻を取得する wall-clock のふたつのインターフェースを定義しています。

リスト 4.43: wasi:clocks/monotonic-clock インターフェース

```
package wasi:clocks@0.2.0;

interface monotonic-clock {
    use wasi:io/poll@0.2.0.{pollable};

    type instant = u64;
    type duration = u64;

    now: func() -> instant;
    resolution: func() -> duration;
    subscribe-instant: func(when: instant) -> pollable;
    subscribe-duration: func(when: duration) -> pollable;
}
```

リスト 4.44: wasi:clocks/wall-clock インターフェース

```
package wasi:clocks@0.2.0;

interface wall-clock {
    record datetime {
        seconds: u64,
        nanoseconds: u32,
    }

    now: func() -> datetime;
```

```
    resolution: func() -> datetime;
}
```

現在時刻 (UNIXTIME) を表示する

wasi:clocks/wall-clock インターフェースを使用して、現在時刻(UNIXTIME)を表示する Wasm
を作成してみましょう。新しい Rust のパッケージを作成し、WIT ファイルを用意します。

```
$ cargo component new --lib clocks_wall
...
```

リスト 4.45: clocks_wall/wit/world.wit

```
package component:clocks-wall;

world example {
    import wasi:clocks/wall-clock@0.2.0;
    import wasi:cli/stdout@0.2.0;
    export wasi:cli/run@0.2.0;
}
```

必要な WIT 定義を追加し、依存関係に追加します。

リスト 4.46: clocks_wall/wit/cli/run.wit

```
package wasi:cli@0.2.0;

interface run {
    run: func() -> result;
}
```

リスト 4.47: clocks_wall/wit/cli/stdio.wit

```
package wasi:cli@0.2.0;

interface stdout {
    use wasi:io/streams@0.2.0.{output-stream};
    get-stdout: func() -> output-stream;
}
```

第 4 章　WASI 0.2 ｜ 93

```
$ cargo component add --path ./wit/cli --target wasi:cli
        Added dependency `wasi:cli` from path `./wit/cli`
```

リスト 4.48: clocks_wall/wit/io/streams.wit

```
package wasi:io@0.2.0;

interface streams {
    use error.{error};
    use poll.{pollable};

...

    resource output-stream {
...
    }
}
```

リスト 4.49: clocks_wall/wit/io/poll.wit

```
package wasi:io@0.2.0;

interface poll {
    resource pollable {
        ready: func() -> bool;
        block: func();
    }

    poll: func(in: list<borrow<pollable>>) -> list<u32>;
}
```

リスト 4.50: clocks_wall/wit/io/error.wit

```
package wasi:io@0.2.0;

interface error {
    resource error {
        to-debug-string: func() -> string;
```

```
        }
}
```

```
$ cargo component add --path ./wit/io --target wasi:io
        Added dependency `wasi:io` from path `./wit/io`
```

リスト4.51: clocks_wall/wit/clocks/wall-clock.wit

```
package wasi:clocks@0.2.0;

interface wall-clock {
    record datetime {
        seconds: u64,
        nanoseconds: u32,
    }


    now: func() -> datetime;
    resolution: func() -> datetime;
}
```

```
$ cargo component add --path ./wit/clocks --target wasi:clocks
        Added dependency `wasi:clocks` from path `./wit/clocks`
```

src/lib.rsを次のように変更し、ビルドすると、clocks_wall.wasmが生成されます。

リスト4.52: clocks_wall/src/lib.rs

```
#[allow(warnings)]
mod bindings;

use bindings::exports::wasi::cli::run::Guest;
use bindings::wasi::cli::stdout::get_stdout;
use bindings::wasi::clocks::wall_clock::now;

struct Component;
impl Guest for Component {
    fn run() -> Result<(), ()> {
        let current_time = now().seconds;
        let stdout = get_stdout();
        stdout
```

第4章　WASI 0.2　95

```
            .blocking_write_and_flush(format!("{current_time}\n").as_bytes())
            .unwrap();

        Ok(())
    }
}


bindings::export!(Component with_types_in bindings);
```

ビルドして実行すると、現在時刻(UNIXTIME)が表示されます。

```
$ cargo component build --release
   Generating bindings for clocks_wall (src/bindings.rs)
    Compiling wit-bindgen-rt v0.27.0
    Compiling bitflags v2.6.0
    Compiling clocks_wall v0.1.0 (/wasi-book-example/第4章/clocks_wall)
     Finished `release` profile [optimized] target(s) in 0.72s
     Creating component target/wasm32-wasi/release/clocks_wall.wasm
```

```
$ wasmtime run target/wasm32-wasi/release/clocks_wall.wasm
1716627600
```

4.5 wasi:randomパッケージ

wasi:randomは、乱数を生成するためのrandomインターフェースとinsecureインターフェース
を定義しています。多くのプログラミング言語ではrandomというと通常は疑似乱数のことを指し、
暗号論的疑似乱数はオプション扱いですが、WASI 0.2ではrandomは暗号論的疑似乱数のことを指
し、通常の疑似乱数はinsecureとして区別しています。

リスト4.53: wasi:random/randomインターフェース

```
package wasi:random@0.2.0;

interface random {
    get-random-bytes: func(len: u64) -> list<u8>;

    get-random-u64: func() -> u64;
}
```

リスト4.54: wasi:random/insecure インターフェース

```
package wasi:random@0.2.0;

interface insecure {
    get-insecure-random-bytes: func(len: u64) -> list<u8>;
    get-insecure-random-u64: func() -> u64;
}
```

乱数を表示する

wasi:random/random インターフェースを使用して、乱数を表示するプログラムを作成してみましょう。新しいRustのパッケージを作成し、WITファイルを用意します。

```
$ cargo component new --lib random
...
```

リスト4.55: random/wit/world.wit

```
package component:random;

world example {
    import wasi:random/random@0.2.0;
    import wasi:cli/stdout@0.2.0;
    export wasi:cli/run@0.2.0;
}
```

必要なWIT定義を追加し、依存関係に追加します。

リスト4.56: random/wit/cli/run.wit

```
package wasi:cli@0.2.0;

interface run {
    run: func() -> result;
}
```

第4章　WASI 0.2　97

リスト4.57: random/wit/cli/stdio.wit

```
package wasi:cli@0.2.0;

interface stdout {
    use wasi:io/streams@0.2.0.{output-stream};
    get-stdout: func() -> output-stream;
}
```

```
$ cargo component add --path ./wit/cli --target wasi:cli
        Added dependency `wasi:cli` from path `./wit/cli`
```

リスト4.58: random/wit/io/streams.wit

```
package wasi:io@0.2.0;

interface streams {
    use error.{error};
    use poll.{pollable};

...

    resource output-stream {
...
    }
}
```

リスト4.59: random/wit/io/poll.wit

```
package wasi:io@0.2.0;

interface poll {
    resource pollable {
        ready: func() -> bool;
        block: func();
    }

    poll: func(in: list<borrow<pollable>>) -> list<u32>;
}
```

リスト 4.60: random/wit/io/error.wit

```
package wasi:io@0.2.0;


interface error {
    resource error {
        to-debug-string: func() -> string;

    }
}
```

```
$ cargo component add --path ./wit/io --target wasi:io
      Added dependency `wasi:io` from path `./wit/io`
```

リスト 4.61: random/wit/random/random.wit

```
package wasi:random@0.2.0;


interface random {
    get-random-bytes: func(len: u64) -> list<u8>;

    get-random-u64: func() -> u64;
}
```

```
$ cargo component add --path ./wit/random --target wasi:random
      Added dependency `wasi:random` from path `./wit/random`
```

src/lib.rsを次のように変更し、ビルドするとrandom.wasmが生成されます。

リスト 4.62: random/src/lib.rs

```
#[allow(warnings)]
mod bindings;

use bindings::exports::wasi::cli::run::Guest;
use bindings::wasi::cli::stdout::get_stdout;
use bindings::wasi::random::random::get_random_u64;


struct Component;
impl Guest for Component {
    fn run() -> Result<(), ()> {
        let random = get_random_u64();
        let stdout = get_stdout();
```

第4章　WASI 0.2 | 99

```
        stdout
            .blocking_write_and_flush(format!("{random}\n").as_bytes())
            .unwrap();

        Ok(())
    }
}

bindings::export!(Component with_types_in bindings);
```

ビルドして実行すると、乱数が表示されます。

```
$ cargo component build --release
   Generating bindings for random (src/bindings.rs)
    Compiling wit-bindgen-rt v0.27.0
    Compiling bitflags v2.6.0
    Compiling random v0.1.0 (/wasi-book-example/第4章/random)
     Finished `release` profile [optimized] target(s) in 0.73s
      Creating component target/wasm32-wasi/release/random.wasm
```

```
$ wasmtime run target/wasm32-wasi/release/random.wasm
4944205001443375959
```

4.6　wasi:filesystemパッケージ

wasi:filesystemパッケージは、ファイルを読み書きするためのインターフェースとしてファイル操作のためのリソースを定義しているtypesインターフェースと、アクセス可能なディレクトリーのリストを返すpreopensインターフェースのふたつを定義しています。

リスト4.63: wasi:filesystem/types インターフェース

```
package wasi:filesystem@0.2.0;

interface types {
    use wasi:io/streams@0.2.0.{input-stream, output-stream, error};
    use wasi:clocks/wall-clock@0.2.0.{datetime};

...

    resource descriptor {
...
```

100 │ 第4章 WASI 0.2

```
        read: func(
            length: filesize,
            offset: filesize,
        ) -> result<tuple<list<u8>, bool>, error-code>;

...

        open-at: func(
            path-flags: path-flags,
            path: string,
            open-flags: open-flags,
            %flags: descriptor-flags,
        ) -> result<descriptor, error-code>;

...
    }

...
}
```

リスト 4.64: wasi:filesystem/preopens インターフェース

```
package wasi:filesystem@0.2.0;

interface preopens {
    use types.{descriptor};

    get-directories: func() -> list<tuple<descriptor, string>>;
}
```

preopens インターフェース

preopens インターフェースは WASI 独自の概念です。WASI は、デフォルトで Wasm がファイルシステムにアクセスすることを制限しています。ファイルにアクセスする場合、ランタイムは preopens インターフェースを通してアクセス可能なディレクトリーを事前に指定します。Wasm のプログラムは、preopens インターフェースを通して得られるディレクトリーのファイルのみにアクセスできます。

WASI 0.1 は、POSIX API をほぼそのまま移植したものでした。UNIX の世界では「すべてはファイルである」という考えのため、標準出力もファイルも区別せず同じように扱うことができます。

第 4 章　WASI 0.2　│　101

そのため、WASI 0.1では fd_write 関数を通してファイルにも標準出力にも同じように書き込むことができましたが、WASI 0.2では標準出力や標準入力は wasi:cli によって提供され、ファイルとは区別されています。

ファイルの内容を表示する

wasi:filesystem パッケージを使用して、ファイルの内容を表示する Wasm を作成してみましょう。新しい Rust のパッケージを作成し、WIT ファイルを用意します。

```
$ cargo component new --lib filesystem
...
```

リスト 4.65: filesystem/wit/world.wit

```
package component:filesystem;

world example {
    import wasi:filesystem/preopens@0.2.0;
    import wasi:cli/stdout@0.2.0;
    import wasi:cli/environment@0.2.0;
    export wasi:cli/run@0.2.0;
}
```

必要な WIT 定義を追加し、依存関係に追加します。

リスト 4.66: filesystem/wit/cli/run.wit

```
package wasi:cli@0.2.0;

interface run {
    run: func() -> result;
}
```

リスト 4.67: filesystem/wit/cli/stdio.wit

```
package wasi:cli@0.2.0;

interface stdout {
    use wasi:io/streams@0.2.0.{output-stream};
    get-stdout: func() -> output-stream;
}
```

102　第4章　WASI 0.2

リスト 4.68: filesystem/wit/cli/environment.wit

```
package wasi:cli@0.2.0;

interface environment {
    get-environment: func() -> list<tuple<string, string>>;
    get-arguments: func() -> list<string>;
    initial-cwd: func() -> option<string>;
}
```

```
$ cargo component add --path ./wit/cli --target wasi:cli
        Added dependency `wasi:cli` from path `./wit/cli`
```

リスト 4.69: filesystem/wit/io/streams.wit

```
package wasi:io@0.2.0;

interface streams {
    use error.{error};
    use poll.{pollable};

...

    resource output-stream {
...
    }
}
```

リスト 4.70: filesystem/wit/io/poll.wit

```
package wasi:io@0.2.0;

interface poll {
    resource pollable {
        ready: func() -> bool;
        block: func();
    }

    poll: func(in: list<borrow<pollable>>) -> list<u32>;
```

```
}
```

リスト 4.71: filesystem/wit/io/error.wit

```
package wasi:io@0.2.0;

interface error {
    resource error {
        to-debug-string: func() -> string;
    }
}
```

```
$ cargo component add --path ./wit/io --target wasi:io
        Added dependency `wasi:io` from path `./wit/io`
```

リスト 4.72: filesystem/wit/clocks/wall-clock.wit

```
package wasi:clocks@0.2.0;

interface wall-clock {
    record datetime {
        seconds: u64,
        nanoseconds: u32,
    }

    now: func() -> datetime;
    resolution: func() -> datetime;
}
```

```
$ cargo component add --path ./wit/clocks --target wasi:clocks
        Added dependency `wasi:clocks` from path `./wit/clocks`
```

リスト 4.73: filesystem/wit/filesystem/preopens.wit

```
package wasi:filesystem@0.2.0;

interface preopens {
```

104 ｜ 第4章 WASI 0.2

```
    use types.{descriptor};

    get-directories: func() -> list<tuple<descriptor, string>>;
}
```

リスト 4.74: filesystem/wit/filesystem/types.wit

```
package wasi:filesystem@0.2.0;
interface types {
    use wasi:io/streams@0.2.0.{input-stream, output-stream, error};
    use wasi:clocks/wall-clock@0.2.0.{datetime};

...

    record metadata-hash-value {
        lower: u64,
        upper: u64,
    }

    resource descriptor {
...

        read: func(
            length: filesize,
            offset: filesize,
        ) -> result<tuple<list<u8>, bool>, error-code>;

...

        open-at: func(
            path-flags: path-flags,
            path: string,
            open-flags: open-flags,
            %flags: descriptor-flags,
        ) -> result<descriptor, error-code>;

...
    }

...
}
```

```
$ cargo component add --path ./wit/filesystem --target wasi:filesystem
        Added dependency `wasi:filesystem` from path `./wit/filesystem`
```

src/lib.rsを次のように変更し、ビルドすると、filesystem.wasmが生成されます。

リスト 4.75: filesystem/src/lib.rs

```rust
#[allow(warnings)]
mod bindings;

use bindings::exports::wasi::cli::run::Guest;
use bindings::wasi::cli::environment::get_arguments;
use bindings::wasi::cli::stdout::get_stdout;
use bindings::wasi::filesystem::preopens::get_directories;
use bindings::wasi::filesystem::types::{DescriptorFlags, OpenFlags, PathFlags};

struct Component;

impl Guest for Component {
    fn run() -> Result<(), ()> {
        let (dir, _) = &get_directories()[0];

        // Open file
        let filename = &get_arguments()[1];
        let file = dir
            .open_at(
                PathFlags::empty(),
                filename,
                OpenFlags::empty(),
                DescriptorFlags::READ,
            )
            .unwrap();

        // Read file
        let (bin, _) = file.read(1024, 0).unwrap();
        let stdout = get_stdout();
        stdout.blocking_write_and_flush(&bin).unwrap();
        Ok(())
    }
}

bindings::export!(Component with_types_in bindings);
```

```
$ cargo component build --release
  Generating bindings for filesystem (src/bindings.rs)
    Compiling wit-bindgen-rt v0.27.0
    Compiling bitflags v2.6.0
    Compiling filesystem v0.1.0 (/wasi-book-example/第4章/filesystem)
     Finished `release` profile [optimized] target(s) in 1.11s
      Creating component target/wasm32-wasi/release/filesystem.wasm
```

実行時に`--dir`オプションでアクセス可能なディレクトリーを指定し、引数に表示するファイル
を指定すると、ファイルの内容が表示されます。

```
$ wasmtime run --dir . \
    target/wasm32-wasi/release/filesystem.wasm Cargo.toml
[package]
name = "filesystem"
version = "0.1.0"
edition = "2021"
...
```

4.7 wasi:sockets パッケージ

wasi:socketsパッケージはソケットのリソースを定義しているtcpインターフェースや
udpインターフェース、新しいソケットを作成するtcp-create-socketインターフェースや
udp-create-socketインターフェースといったソケット通信を行うためのインターフェースを
定義しています。WASI 0.1ではソケットの作成はサポートされていませんでしたが、WASI 0.2では
tcp-create-socketインターフェースとudp-create-socketインターフェースのふたつがソケット
を作成するインターフェースとして定義されています。

リスト4.76: wasi:sockets/tcp インターフェース

```
package wasi:sockets@0.2.0;

interface tcp {
    use wasi:io/streams@0.2.0.{input-stream, output-stream};
    use wasi:io/poll@0.2.0.{pollable};
    use wasi:clocks/monotonic-clock@0.2.0.{duration};
    use network.{network, error-code, ip-socket-address, ip-address-family};

    ...

    resource tcp-socket {
        start-bind: func(
```

```
        network: borrow<network>,
        local-address: ip-socket-address,
    ) -> result<_, error-code>;
    finish-bind: func() -> result<_, error-code>;

    start-connect: func(
        network: borrow<network>,
        remote-address: ip-socket-address,
    ) -> result<_, error-code>;
    finish-connect: func() -> result<
        tuple<input-stream, output-stream>,
        error-code
    >;

    start-listen: func() -> result<_, error-code>;
    finish-listen: func() -> result<_, error-code>;

    accept: func() -> result<
        tuple<tcp-socket, input-stream, output-stream>,
        error-code
    >;

...

    subscribe: func() -> pollable;

    shutdown: func(shutdown-type: shutdown-type) -> result<_, error-code>;
    }
}
```

リスト4.77: wasi:sockets/udp インターフェース

```
package wasi:sockets@0.2.0;

interface udp {
    use wasi:io/poll@0.2.0.{pollable};
    use network.{network, error-code, ip-socket-address, ip-address-family};

...
```

```
    resource udp-socket {
...
    }


...
}
```

リスト4.78: wasi:sockets/tcp-create-socket インターフェース

```
package wasi:sockets@0.2.0;

interface tcp-create-socket {
    use network.{network, error-code, ip-address-family};
    use tcp.{tcp-socket};

    create-tcp-socket: func(
        address-family: ip-address-family,
    ) -> result<tcp-socket, error-code>;
}
```

リスト4.79: wasi:sockets/udp-create-socket インターフェース

```
interface udp-create-socket {
    use network.{network, error-code, ip-address-family};
    use udp.{udp-socket};

    create-udp-socket: func(
        address-family: ip-address-family,
    ) -> result<udp-socket, error-code>;
}
```

HTTPサーバーを作成する

wasi:socketsパッケージを使用して、HTTPリクエストのホスティングを行うWasmを作成してみましょう。新しいRustのパッケージを作成し、WITファイルを用意します。

```
$ cargo component new --lib sockets_tcp
...
```

第4章　WASI 0.2 | 109

リスト4.80: sockets_tcp/wit/world.wit

```
package component:sockets-tcp;

world example {
    import wasi:sockets/instance-network@0.2.0;
    import wasi:sockets/tcp-create-socket@0.2.0;
    import wasi:cli/stdout@0.2.0;
    export wasi:cli/run@0.2.0;
}
```

必要なWIT定義を追加し、依存関係に追加します。

リスト4.81: sockets_tcp/wit/cli/run.wit

```
package wasi:cli@0.2.0;

interface run {
    run: func() -> result;
}
```

リスト4.82: sockets_tcp/wit/cli/stdio.wit

```
package wasi:cli@0.2.0;

interface stdout {
    use wasi:io/streams@0.2.0.{output-stream};
    get-stdout: func() -> output-stream;
}
```

```
$ cargo component add --path ./wit/cli --target wasi:cli
        Added dependency `wasi:cli` from path `./wit/cli`
```

リスト4.83: sockets_tcp/wit/io/streams.wit

```
package wasi:io@0.2.0;

interface streams {
    use error.{error};
    use poll.{pollable};
```

110 | 第4章 WASI 0.2

```
...

    resource output-stream {
...

    }
}
```

リスト 4.84: sockets_tcp/wit/io/poll.wit

```
package wasi:io@0.2.0;

interface poll {
    resource pollable {
        ready: func() -> bool;
        block: func();
    }

    poll: func(in: list<borrow<pollable>>) -> list<u32>;
}
```

リスト 4.85: sockets_tcp/wit/io/error.wit

```
package wasi:io@0.2.0;

interface error {
    resource error {
        to-debug-string: func() -> string;
    }
}
```

```
$ cargo component add --path ./wit/io --target wasi:io
        Added dependency `wasi:io` from path `./wit/io`
```

リスト 4.86: sockets_tcp/wit/clocks/monotonic-clock.wit

```
package wasi:clocks@0.2.0;

interface monotonic-clock {
```

第4章 WASI 0.2 | 111

```
    use wasi:io/poll@0.2.0.{pollable};

    type instant = u64;
    type duration = u64;

    now: func() -> instant;
    resolution: func() -> duration;
    subscribe-instant: func(when: instant) -> pollable;
    subscribe-duration: func(when: duration) -> pollable;
}
```

```
$ cargo component add --path ./wit/clocks --target wasi:clocks
      Added dependency `wasi:clocks` from path `./wit/clocks`
```

リスト 4.87: sockets_tcp/wit/sockets/network.wit

```
package wasi:sockets@0.2.0;

interface network {
    resource network;

...
}
```

リスト 4.88: sockets_tcp/wit/sockets/instance-network.wit

```
package wasi:sockets@0.2.0;

interface instance-network {
    use network.{network};

    instance-network: func() -> network;
}
```

112 | 第4章 WASI 0.2

リスト4.89: sockets_tcp/wit/sockets/tcp.wit

```
package wasi:sockets@0.2.0;

interface tcp {
    use wasi:io/streams@0.2.0.{input-stream, output-stream};
    use wasi:io/poll@0.2.0.{pollable};
    use wasi:clocks/monotonic-clock@0.2.0.{duration};
    use network.{network, error-code, ip-socket-address, ip-address-family};

...


    resource tcp-socket {
...
    }
}
```

リスト4.90: sockets_tcp/wit/sockets/tcp-create-socket.wit

```
package wasi:sockets@0.2.0;

interface tcp-create-socket {
    use network.{network, error-code, ip-address-family};
    use tcp.{tcp-socket};

    create-tcp-socket: func(
        address-family: ip-address-family,
    ) -> result<tcp-socket, error-code>;
}
```

```
$ cargo component add --path ./wit/sockets --target wasi:sockets
      Added dependency `wasi:sockets` from path `./wit/sockets`
```

src/lib.rsを次のように変更し、ビルドすると、sockets_tcp.wasmが生成されます。

リスト4.91: sockets_tcp/src/lib.rs

```
#[allow(warnings)]
mod bindings;

use bindings::exports::wasi::cli::run::Guest;
use bindings::wasi::cli::stdout::get_stdout;
```

第4章　WASI 0.2 ｜ 113

```rust
use bindings::wasi::sockets::instance_network::instance_network;
use bindings::wasi::sockets::network::{IpAddressFamily, IpSocketAddress,
Ipv4SocketAddress};
use bindings::wasi::sockets::tcp::ShutdownType;
use bindings::wasi::sockets::tcp_create_socket::create_tcp_socket;

struct Component;

const HTTP_RESPONSE: &str =
    "HTTP/1.1 200 OK\r\nContent-Type: text/plain\r\nContent-Length: 22\n\nHello,
wasi:sockets!\r\n";

impl Guest for Component {
    fn run() -> Result<(), ()> {
        let stdout = get_stdout();
        let socket = create_tcp_socket(IpAddressFamily::Ipv4).unwrap();

        let network = instance_network();
        let address = IpSocketAddress::Ipv4(Ipv4SocketAddress {
            port: 8080,
            address: (127, 0, 0, 1),
        });

        // await bind
        stdout.blocking_write_and_flush(b"start_bind...").unwrap();
        socket.start_bind(&network, address).unwrap();
        socket.subscribe().block();

        stdout.blocking_write_and_flush(b"OK!\n").unwrap();
        socket.finish_bind().unwrap();

        // await listen
        stdout.blocking_write_and_flush(b"start_listen...").unwrap();
        socket.start_listen().unwrap();
        socket.subscribe().block();

        stdout.blocking_write_and_flush(b"OK!\n").unwrap();
        socket.finish_listen().unwrap();

        // await accept
        stdout.blocking_write_and_flush(b"accepting...").unwrap();
```

```
        socket.subscribe().block();

        stdout.blocking_write_and_flush(b"Accepted!\n").unwrap();
        let (socket, _, output) = socket.accept().unwrap();
        output
            .blocking_write_and_flush(HTTP_RESPONSE.as_bytes())
            .unwrap();
        drop(output);

        stdout
            .blocking_write_and_flush(b"socket shutdown...\n")
            .unwrap();
        socket.shutdown(ShutdownType::Both).unwrap();
        drop(socket);
        Ok(())
    }
}

bindings::export!(Component with_types_in bindings);
```

```
$ cargo component build --release
  Generating bindings for sockets_tcp (src/bindings.rs)
   Compiling sockets_tcp v0.1.0 (/wasi-book-example/第4章/sockets_tcp)
    Finished `release` profile [optimized] target(s) in 0.30s
    Creating component target/wasm32-wasi/release/sockets_tcp.wasm
```

このWasmはTCPソケットを作成して、8080番ポートでリクエストを待ち受けます。リクエスト
を受け取ると、レスポンスを返してソケットを閉じます。Wasmtimeはネットワークをデフォルト
で利用できないようになっているため、実行時に-S inherit-networkオプションを指定してネッ
トワークを利用できるようにします。

```
$ wasmtime run -S inherit-network \
    ./target/wasm32-wasi/release/sockets_tcp.wasm
start_bind...OK!
start_listen...OK!
accepting...
```

実行すると、クライアントからのリクエストを待ち受けている状態になります。この状態で別のター
ミナルを開き、curlコマンドでリクエストを送信すると、「Hello, wasi:sockets!」というレスポンス
が返ってきます。

第4章 WASI 0.2 | 115

```
$ curl http://localhost:8080
Hello, wasi:sockets!
```

サーバーはレスポンスを返した後、TCPソケットを閉じています。

```
$ wasmtime run -S=inherit-network \
    ./target/wasm32-wasi/release/sockets_tcp.wasm
start_bind...OK!
start_listen...OK!
accepting...Accepted!
socket shutdown...
```

このようにWASI 0.2ではソケットの作成がサポートされたため、Wasmコンポーネントだけで
HTTPサーバーを実装できるようになりました。

4.8　wasi:httpパッケージ

wasi:httpパッケージは、HTTPリクエストを受信するためのincoming-handlerインターフェー
スと、HTTPリクエストを送信するためのoutgoing-handlerインターフェースが定義されていま
す。すでにネットワーク通信を行うためのwasi:socketsパッケージがありますが、wasi:httpパッ
ケージはHTTPの通信に特化した高レイヤーのインターフェースを提供します。

リスト4.92: wasi:http/incoming-handlerインターフェース

```
package wasi:http@0.2.0;

interface incoming-handler {
    use types.{incoming-request, response-outparam};

    handle: func(request: incoming-request, response-out: response-outparam);
}
```

リスト4.93: wasi:http/outgoing-handlerインターフェース

```
interface outgoing-handler {
    use types.{
        outgoing-request, request-options, future-incoming-response, error-code
    };

    handle: func(
        request: outgoing-request,
```

```
        options: option<request-options>
    ) -> result<future-incoming-response, error-code>;
}
```

wasi:http/proxy ワールド

wasi:http/proxy ワールドは wasi:http/incoming-handler インターフェースをエクスポートしており、wasi:cli/command ワールドの他に WASI 0.2 で定義されているもうひとつのエントリーポイントとなります。

このワールド自体は、直接ソケットを扱うといったことはしません。wasi:http/proxy ワールドは、エッジコンピューティング(サーバーレス環境)における HTTP リクエストのエントリーポイントとして利用することを想定したワールドです。

リスト 4.94: wasi:http/proxy ワールド

```
package wasi:http@0.2.0;

world proxy {
    include wasi:clocks/imports@0.2.0;
    import wasi:random/random@0.2.0;
    import wasi:cli/stdout@0.2.0;
    import wasi:cli/stderr@0.2.0;
    import wasi:cli/stdin@0.2.0;
    import outgoing-handler;

    export incoming-handler;
}
```

HTTP サーバーを作成する

wasi:http/proxy ワールドを用いて、HTTP リクエストをホスティングする Wasm を作成してみましょう。新しい Rust のパッケージを作成し、WIT ファイルを用意します。

```
$ cargo component new --lib http_proxy
...
```

第 4 章　WASI 0.2　｜　117

リスト 4.95: http_proxy/wit/world.wit

```
package component:http-proxy;

world example {
    include wasi:http/proxy@0.2.0;
}
```

必要な WIT 定義を追加し、依存関係に追加します。

リスト 4.96: http_proxy/wit/http/proxy.wit

```
package wasi:http@0.2.0;

world proxy {
    include wasi:clocks/imports@0.2.0;
    import wasi:random/random@0.2.0;
    import wasi:cli/stdout@0.2.0;
    import wasi:cli/stderr@0.2.0;
    import wasi:cli/stdin@0.2.0;
    import outgoing-handler;

    export incoming-handler;
}
```

リスト 4.97: http_proxy/wit/http/handler.wit

```
package wasi:http@0.2.0;

interface incoming-handler {
    use types.{incoming-request, response-outparam};

    handle: func(
        request: incoming-request,
        response-out: response-outparam
    );
}

interface outgoing-handler {
    use types.{
        outgoing-request, request-options, future-incoming-response, error-code
    };
```

118 | 第 4 章 WASI 0.2

```
    handle: func(
        request: outgoing-request,
        options: option<request-options>
    ) -> result<future-incoming-response, error-code>;
}
```

リスト 4.98: http_proxy/wit/http/types.wit

```
package wasi:http@0.2.0;

interface types {
    use wasi:clocks/monotonic-clock@0.2.0.{duration};
    use wasi:io/streams@0.2.0.{input-stream, output-stream};
    use wasi:io/error@0.2.0.{error as io-error};
    use wasi:io/poll@0.2.0.{pollable};

...
}
```

```
$ cargo component add --path ./wit/http --target wasi:http
      Added dependency `wasi:http` from path `./wit/http`
```

リスト 4.99: http_proxy/wit/clocks/imports.wit

```
package wasi:clocks@0.2.0;

world imports {
    import monotonic-clock;
    import wall-clock;
}
```

リスト 4.100: http_proxy/wit/clocks/monotonic-clock.wit

```
package wasi:clocks@0.2.0;

interface monotonic-clock {
    use wasi:io/poll@0.2.0.{pollable};
```

第4章 WASI 0.2 119

```
    type instant = u64;
    type duration = u64;

    now: func() -> instant;
    resolution: func() -> duration;
    subscribe-instant: func(when: instant) -> pollable;
    subscribe-duration: func(when: duration) -> pollable;
}
```

リスト4.101: http_proxy/wit/clocks/wall-clock.wit

```
package wasi:clocks@0.2.0;

interface wall-clock {
    record datetime {
        seconds: u64,
        nanoseconds: u32,
    }

    now: func() -> datetime;
    resolution: func() -> datetime;
}
```

```
$ cargo component add --path ./wit/clocks --target wasi:clocks
      Added dependency `wasi:clocks` from path `./wit/clocks`
```

リスト4.102: http_proxy/wit/io/streams.wit

```
package wasi:io@0.2.0;

interface streams {
    use error.{error};
    use poll.{pollable};

    ...
}
```

120 │ 第4章 WASI 0.2

リスト 4.103: http_proxy/wit/io/poll.wit

```
package wasi:io@0.2.0;

interface poll {
    resource pollable {
        ready: func() -> bool;
        block: func();
    }


    poll: func(in: list<borrow<pollable>>) -> list<u32>;
}
```

リスト 4.104: http_proxy/wit/io/error.wit

```
package wasi:io@0.2.0;

interface error {
    resource error {
        to-debug-string: func() -> string;
    }
}
```

```
$ cargo component add --path ./wit/io --target wasi:io
      Added dependency `wasi:io` from path `./wit/io`
```

リスト 4.105: http_proxy/wit/random/random.wit

```
package wasi:random@0.2.0;

interface random {
    get-random-bytes: func(len: u64) -> list<u8>;

    get-random-u64: func() -> u64;
}
```

```
$ cargo component add --path ./wit/random --target wasi:random
      Added dependency `wasi:random` from path `./wit/random`
```

第4章　WASI 0.2　121

リスト 4.106: http_proxy/wit/cli/stdio.wit

```
package wasi:cli@0.2.0;

interface stdin {
    use wasi:io/streams@0.2.0.{input-stream};
    get-stdin: func() -> input-stream;
}

interface stdout {
    use wasi:io/streams@0.2.0.{output-stream};
    get-stdout: func() -> output-stream;
}

interface stderr {
    use wasi:io/streams@0.2.0.{output-stream};
    get-stderr: func() -> output-stream;
}
```

```
$ cargo component add --path ./wit/cli --target wasi:cli
        Added dependency `wasi:cli` from path `./wit/cli`
```

src/lib.rsを次のように変更します。

リスト 4.107: http_proxy/src/lib.rs

```
#[allow(warnings)]
mod bindings;

use crate::bindings::exports::wasi::http::incoming_handler::Guest;
use bindings::wasi::http::types::{
    Headers, IncomingRequest, OutgoingBody, OutgoingResponse, ResponseOutparam,
};

struct Component;

impl Guest for Component {
    fn handle(_: IncomingRequest, outparam: ResponseOutparam) {
        let resp = OutgoingResponse::new(Headers::new());

        let body = resp.body().unwrap();
        let output = body.write().unwrap();
        output
```

```
            .blocking_write_and_flush(b"Hello, wasi:http/proxy!\n")
            .unwrap();
        drop(output);
        OutgoingBody::finish(body, None).unwrap();

        ResponseOutparam::set(outparam, Ok(resp));
    }
}


bindings::export!(Component with_types_in bindings);
```

Cargo.tomlに package.metadata.component.proxy = trueを追加し、ビルドします。

リスト4.108: http_proxy/Cargo.toml (一部)

```
...

[package.metadata.component]
package = "component:http-proxy"
proxy = true

...
```

```
$ cargo component build --release
  Generating bindings for http_proxy (src/bindings.rs)
   Compiling wit-bindgen-rt v0.27.0
   Compiling bitflags v2.6.0
   Compiling http_proxy v0.1.0 (/wasi-book-example/第4章/http_proxy)
    Finished `release` profile [optimized] target(s) in 0.84s
    Creating component target/wasm32-wasi/release/http_proxy.wasm
```

wasi:http/proxyワールドの実装は、wasmtime serveコマンドで実行することができます。

```
$ wasmtime serve ./target/wasm32-wasi/release/http_proxy.wasm
Serving HTTP on http://0.0.0.0:8080/
```

別のターミナルを開き、curlコマンドでリクエストを送信すると、レスポンスが返ってくることが
確認できます。

```
$ curl http://localhost:8080
Hello, wasi:http/proxy!
```

4.9 WASI 0.2の課題

WASI 0.2では、WASI 0.1で課題となっていたソケットの作成がサポートされました。WASI 0.2で解決されていない課題として、非同期処理のサポートがあります。非同期処理はWASI 0.3(プレビュー3)でのサポートが期待されています。

非同期処理

WASI 0.3(プレビュー3)では、コンポーネントモデルにfutureとstreamのふたつのキーワードが追加される可能性があります[2]。これらの追加により、async/await構文を扱えるようになることが期待されています。

リスト4.109: futureキーワードの例

```
interface example {
    foo: func() -> future<string>;
}
```

リスト4.110: streamキーワードの例

```
interface example {
    bar: func() -> stream<u8>;
}
```

また、これらのキーワード追加に合わせて、Wasmのコアスペックに複数のスタックを扱えるようにする仕様が追加される可能性があります[3]。

WASI 0.2では非同期処理を扱えないため、wasi:io/pollインターフェースのpollableを用いて実行待ちを行っています。wasi:socketsパッケージのTCPソケットでは次のようにpollableを返すsubscribe関数が用意されており、次のようにソケットの受信待ちを行います。

リスト4.111: WASI 0.2のソケット実装例

```
let socket = create_tcp_socket(IpAddressFamily::Ipv4).unwrap();

// await bind
socket.start_bind(&network, address).unwrap();
socket.subscribe().block();
socket.finish_bind().unwrap();

// await listen
```

2.https://docs.google.com/presentation/d/1MNVOZ8hdofO3tl0szg_i-Yoy0N2QPU2C--LzVuoGSIE

3.stack-swiching :https://github.com/WebAssembly/stack-switching

124　　第4章　WASI 0.2

```
socket.start_listen().unwrap();
socket.subscribe().block();
socket.finish_listen().unwrap();

// await accept
socket.subscribe().block();
let (socket, _, output) = socket.accept().unwrap();
```

WASI 0.3ではコンポーネントモデルにfuture構文が入ることで、pollableを用いずにasync/await構文を用いたインターフェース仕様に修正される可能性があります。

リスト4.112: WASI 0.3で期待されるソケット実装例

```
let socket = create_tcp_socket(IpAddressFamily::Ipv4).unwrap();

// await bind
socket.bind(&network, address).await.unwrap();

// await listen
socket.listen().await.unwrap();

// await accept
let (socket, _, output) = socket.accept().await.unwrap();
```

4.10 まとめ

　WASI 0.2ではコンポーネントモデルを用いることで機能ごとに複数のパッケージに分けられ、C言語ライクなAPIから脱却しました。WASI 0.1で課題となっていたソケットの作成がサポートされたほか、HTTPサーバー用のエントリーポイントとしてwasi:http/proxyワールドが追加されたりと、WASI 0.1と比べて使いやすいインターフェースとなりました。

　WASI 0.2では非同期処理がサポートされていませんが、WASI 0.3でサポートされることが期待されています。

　WASIの仕様は0.3に向けて更新されているため、今後も新しいインターフェースやパッケージが追加されます。気になる方は、WASIのプロポーザル一覧[4]を参照してみてください。

4.WASIプロポーザル一覧 :https://github.com/WebAssembly/WASI/blob/main/Proposals.md

第4章　WASI 0.2　｜　125

第5章 今後の展望

WASI 0.3の開発以外にも、コンポーネントモデルを利用するためのエコシステムの開発が進められています。

5.1 Warg

Warg(WebAssembly Registry)[1]はWasmコンポーネントのためのレジストリで、WITやWasmコンポーネントのアップロードやダウンロードを行うことができます。

現在はまだ開発段階ですが、パブリックベータ版のhttps://wa.dev/が公開されており、WASI 0.2の各コンポーネントのWITパッケージ[2]の仕様を見たりすることができます。

wargコマンドのインストール

wargコマンドは、Rustのパッケージマネージャーであるcargoを用いてインストールすることができます。

```
$ cargo install warg-cli
...
```

wa.devにログインする

https://wa.devにコンポーネントをアップロードするには、ログインしてサーバーにキーを設定する必要があります。キーを設定するには、warg loginコマンドを実行します。

```
$ warg login --registry $YOURNAME.wa.dev
...
```

キーの設定は、Account credentials[3]ページから行うことができます。wargコマンドからのログインする方法もこのページに記載されているため、参考にしてください。

ネームスペースを作成する

ログインしてキーを設定できたら、第3章で作成したhelloコンポーネントとcomponentコンポー

1.https://github.com/bytecodealliance/registry

2.https://wa.dev/wasi

3.Account credentials:https://wa.dev/account/credentials/new

ネントをwa.devにアップロードしてみましょう。

wa.devにアップロードするためには、ネームスペースをwa.dev上に作成する必要があります。ネームスペースの作成は、wa.devのCreate New namespace[4]から行うことができます。ネームスペースはユーザー名などを設定するといいでしょう。

パッケージを作成する

ネームスペースを作成したら、パッケージを作成します。warg publish initコマンドを利用するか、wa.dev上のCreate New package[5]から行うことができます。

```
$ warg publish init asuka:hello
...
```

```
$ warg publish init asuka:component
...
```

コマンドの例では、ネームスペースasukaに対してhelloパッケージとcomponentパッケージを作成しています。

コンポーネントを公開する

パッケージを作成したら、コンポーネントをアップロード(公開)しましょう。wa.devにアップロードするために、WITのパッケージ名とRustの実装を作成したパッケージ名になるよう修正します。

リスト5.1: hello/wit/world.wit
```
package asuka:hello;

interface hello {
    greet: func(name: string) -> string;
}

world greetor {
    export hello;
}
```

4.Create New namespace:https://wa.dev/new/namespace

5.Create New package:https://wa.dev/new

リスト5.2: hello/src/lib.rs

```rust
#[allow(warnings)]
mod bindings;

use crate::bindings::exports::asuka::hello::hello::Guest;

struct Component;

impl Guest for Component {
    fn greet(name: String) -> String {
        format!("Hello, {name}!")
    }
}

bindings::export!(Component with_types_in bindings);
```

コードの修正が完了したら、ビルドしてコンポーネントをアップロードします。コンポーネントの
アップロードには、warg publish releaseコマンドを利用します。

```
$ cargo component build --release
...
```

```
$ warg publish release --name asuka:hello --version 0.1.0 \
    target/wasm32-wasi/release/hello.wasm
...
```

wa.devを開くと、コンポーネントがアップロードされていることを確認できます。component コ
ンポーネントも同様に修正することで、アップロードすることができます。

リスト5.3: component/wit/world.wit

```
package asuka:component;

world hello {
    import asuka:hello/hello;
}
```

128 | 第5章 今後の展望

リスト5.4: component/src/main.rs

```rust
#[allow(warnings)]
mod bindings;

use bindings::askua::hello::hello::greet;

fn main() {
    let greet = greet("asuka");
    println!("{greet}");
}
```

```
$ cargo component build --release
...
```

```
$ warg publish release --name asuka:component --version 0.1.0 \
    target/wasm32-wasi/release/component.wasm
...
```

5.2　WAC

Wasm コンポーネントのインターフェース定義として、WIT(WebAssembly Interface Type)を紹介しました。現在、WITの他にWAC(WebAssembly Compositions)[6]の開発も進められています。

WACはWITのスーパーセットとなる言語とツールとして設計されており、WITにはなかったlet構文などがWACには追加されています。

wac コマンドのインストール

wac コマンドはRustのパッケージマネージャーであるcargoを用いて、インストールすることができます。

```
$ cargo install wac-cli
...
```

コンポーネントを組み合わせる

現在wasm-toolsを用いてコンポーネントを組み合わせることは非推奨となり、代わりにwac コマ

6.https://github.com/bytecodealliance/wac

第5章　今後の展望　| 　129

ンドを用いて複数のWasmコンポーネントを組み合わせることが推奨されています。

```
$ wac plug component/target/wasm32-wasi/release/component.wasm \
    --plug hello/target/wasm32-wasi/release/hello.wasm \
    -o compose.wasm
```

　また、コマンドラインだけでなくWAC言語を用いて、コンポーネントを組み合わせることができます。次のようなWACファイルを作成し、wac コマンドを用いて実行用のexample.wasmを作成することができます。

リスト5.5: example.wac

```
package example:composition;

import environment: wasi:cli/environment@0.2.0;
import exit: wasi:cli/exit@0.2.0;
import error: wasi:io/error@0.2.0;
import streams: wasi:io/streams@0.2.0;
import stdin: wasi:cli/stdin@0.2.0;
import stdout: wasi:cli/stdout@0.2.0;
import stderr: wasi:cli/stderr@0.2.0;
import wall-clock: wasi:clocks/wall-clock@0.2.0;
import types: wasi:filesystem/types@0.2.0;
import preopens: wasi:filesystem/preopens@0.2.0;

let hello = new asuka:hello@0.1.0 {
    environment,
    exit,
    error,
    streams,
    stdin,
    stdout,
    stderr,
    wall-clock,
    types,
    preopens,
};

let instance = new asuka:component@0.1.0 {
    hello: hello.hello,
    environment,
    exit,
```

130 ｜ 第5章　今後の展望

```
        error,
        streams,
        stdin,
        stdout,
        stderr,
        wall-clock,
        types,
        preopens,
    };

    export instance.run;
```

このWACファイルでインポートしているパッケージ名の解決は、Wargサーバー(wa.dev)を通して行われます。

```
$ wac encode -o example.wasm example.wac
    Updating package logs from the registry
 Downloading package content from the registry
  Downloaded package `asuka:hello` 0.1.0
  Downloaded package `wasi:io` 0.2.0
  Downloaded package `wasi:clocks` 0.2.0
  Downloaded package `wasi:cli` 0.2.0
  Downloaded package `wasi:filesystem` 0.2.0
  Downloaded package `asuka:component` 0.1.0
```

```
$ wasmtime run example.wasm
Hello, asuka
```

5.3 WIT定義を公開する

warg publishコマンドでは、コンポーネントを公開することができました。Wargにはコンポーネントだけでなく、WIT定義も公開することができます。

WIT定義を公開するには、witコマンド[7]を利用します。witコマンドは、cargoを用いてインストールすることができます。

7.https://crates.io/crates/wit

第5章　今後の展望 ｜ 131

```
$ cargo install wit
...
```

WIT定義を公開するには、WITファイルのあるディレクトリーで wit init を実行し、wit.toml ファイルを作成します。

```
$ wit init
...
```

依存するパッケージがある場合は、wit add コマンドを使用して追加します。

```
$ wit add wasi:cli
...
```

また、wit.tomlには、公開するWITのバージョン情報などを記述できます。公開する準備ができたら、wit publish コマンドを実行してWIT定義を公開します。

```
$ wit publish --init
...
```

5.4 JCO

JSOは、WasmコンポーネントをJavaScriptから扱えるようにしたりするためのコマンドツールです。

jcoコマンドのインストール

jcoコマンドは、Node.jsのパッケージマネージャーであるnpmを用いてインストールすることができます。

```
$ npm install --global @bytecodealliance/jco
```

コンポーネントをトランスパイルする

WasmコンポーネントをJavaScriptから扱えるようにするには、jcoコマンドを用いてトランスパイルします。

```
$ jco transpile example.wasm -o example-js

  Transpiled JS Component Files:

 - example-js/example.core.wasm                    1.59 MiB
 - example-js/example.core2.wasm                   16.7 KiB
 - example-js/example.core3.wasm                   1.61 MiB
 - example-js/example.core4.wasm                   16.8 KiB
 - example-js/example.d.ts                         0.83 KiB
 - example-js/example.js                           92.1 KiB
...
```

トランスパイルしたコンポーネントを実行する

　トランスパイルができたら、実行用のexample.jsを作成します。

リスト5.6: example.js

```
import { run } from "./example-js/example.js";


run.run();
```

　実行すると、Wasmtimeを用いてexample.wasmを実行したときと同様の結果を得ることができます。

```
$ deno run --allow-read example.js
Hello, asuka
```

5.5　まとめ

　コンポーネントモデルは、異なる言語で実装したWasmであっても、組み合わせてアプリケーションを構築することができる仕組みです。現在、コンポーネントモデルを利用するためのエコシステムとして、レジストリであるWarg、Wasmコンポーネントを組み合わせるためのWAC、JSOなどのツールの開発が進められています。

　これらのエコシステムが出揃ってくると、今後Wasmコンポーネントをより簡単に扱えるようになるでしょう。

付録A　APPENDIX

A.1　参考文献

サンプルコード

https://github.com/a-skua/wasi-book-example

WASI 0.2 Launched

https://bytecodealliance.org/articles/WASI-0.2

Go 1.21 Release Notes

https://tip.golang.org/doc/go1.21

wasm-tools

https://github.com/bytecodealliance/wasm-tools

Wasmtime

https://github.com/bytecodealliance/wasmtimehttps://docs.wasmtime.dev

Rust

https://www.rust-lang.org

cargo-component

https://github.com/bytecodealliance/cargo-component

WebAssembly Text Format Specification

https://webassembly.github.io/spec/core/text/index.html

WebAssembly Spesification

https://webassembly.github.io/spec/core/

WASI 0.1 Application ABI

https://github.com/WebAssembly/WASI/blob/main/legacy/application-abi.md

WASI 0.1 Specification

https://github.com/WebAssembly/WASI/blob/main/legacy/preview1/docs.md

WASI support in Go

https://go.dev/blog/wasi

WASI crate

https://crates.io/crates/wasi

The WebAssembly Component Model

https://component-model.bytecodealliance.org

セマンティックバージョニング

https://semver.org/lang/ja/

Rust Blog Changes to Rust's WASI targets

134 　付録A　APPENDIX

https://blog.rust-lang.org/2024/04/09/updates-to-rusts-wasi-targets.html

Component Model async proposal

https://docs.google.com/presentation/d/1MNVOZ8hdofO3tI0szg_i-Yoy0N2QPU2C--LzVuoGSlE

Stack-Switching Proposal for WebAssembly

https://github.com/WebAssembly/stack-switching

WASI proposals

https://github.com/WebAssembly/WASI/blob/main/Proposals.md

WebAssembly Compositions (WAC)

https://github.com/bytecodealliance/wac

jco

https://github.com/bytecodealliance/jco

著者紹介

八木 明日香 (やぎ あすか)

ソフトウェアエンジニア。2017年よりアルバイト、2020年より会社員としてプログラミングに従事。ウェブシステムの設計から実装、運用、プロジェクト進行まで一通り経験し、クロスプラットフォームを模索していた際にWebAssemblyに出会う。

◎本書スタッフ
アートディレクター/装丁：岡田章志＋GY
編集協力：山部沙織
ディレクター：栗原 翔
〈表紙イラスト〉
桜葉さこく
オンラインゲームのイラストレーターとして数年勤めた後フリーに。ゲーム、書籍、広告のイラストやVTuberなどのキャラクターデザインの他に、UIやwebなどのデザインも行う。自身の作品にヤンデレキッチンシリーズなど。

技術の泉シリーズ・刊行によせて
技術者の知見のアウトプットである技術同人誌は、急速に認知度を高めています。インプレス NextPublishingは国内最大級の即売会「技術書典」（https://techbookfest.org/）で頒布された技術同人誌を底本とした商業書籍を2016年より刊行し、これらを中心とした『技術書典シリーズ』を展開してきました。2019年4月、より幅広い技術同人誌を対象とし、最新の知見を発信するために『技術の泉シリーズ』へリニューアルしました。今後は「技術書典」をはじめとした各種即売会や、勉強会・LT会などで頒布された技術同人誌を底本とした商業書籍を刊行し、技術同人誌の普及と発展に貢献することを目指します。エンジニアの"知の結晶"である技術同人誌の世界に、より多くの方が触れていただくきっかけになれば幸いです。

インプレス NextPublishing
技術の泉シリーズ 編集長 山城 敬

●お断り
掲載したURLは2024年9月1日現在のものです。サイトの都合で変更されることがあります。また、電子版ではURLにハイパーリンクを設定していますが、端末やビューアー、リンク先のファイルタイプによっては表示されないことがあります。あらかじめご了承ください。
●本書の内容についてのお問い合わせ先
株式会社インプレス
インプレス NextPublishing　メール窓口
np-info@impress.co.jp
お問い合わせの際は、書名、ISBN、お名前、お電話番号、メールアドレス に加えて、「該当するページ」と「具体的なご質問内容」「お使いの動作環境」を必ずご明記ください。なお、本書の範囲を超えるご質問にはお答えできないのでご了承ください。
電話やFAXでのご質問には対応しておりません。また、封書でのお問い合わせは回答までに日数をいただく場合があります。あらかじめご了承ください。

●落丁・乱丁本はお手数ですが、インプレスカスタマーセンターまでお送りください。送料弊社負担 てお取り替えさせていただきます。但し、古書店で購入されたものについてはお取り替えできません。
■読者の窓口
インプレスカスタマーセンター
〒101-0051
東京都千代田区神田神保町一丁目105番地
info@impress.co.jp

技術の泉シリーズ
WebAssembly System Interface入門

2024年9月20日　初版発行Ver.1.0（PDF版）

著　者	八木 明日香
編集人	山城 敬
企画・編集	合同会社技術の泉出版
発行人	高橋 隆志
発　行	インプレス NextPublishing
	〒101-0051
	東京都千代田区神田神保町一丁目105番地
	https://nextpublishing.jp/
販　売	株式会社インプレス
	〒101-0051　東京都千代田区神田神保町一丁目105番地

●本書は著作権法上の保護を受けています。本書の一部あるいは全部について株式会社インプレスから文書による許諾を得ずに、いかなる方法においても無断で複写、複製することは禁じられています。

©2024 Asuka Yagi. All rights reserved.
印刷・製本　京葉流通倉庫株式会社
Printed in Japan

ISBN978-4-295-60333-7

●インプレス NextPublishingは、株式会社インプレスR&Dが開発したデジタルファースト型の出版モデルを承継し、幅広い出版企画を電子書籍＋オンデマンドによりスピーディで持続可能な形で実現しています。https://nextpublishing.jp/